CRACKING
THE
CANADIAN
FORMULA

The making of the
Energy and Chemical Workers Union

CRACKING
THE
CANADIAN
FORMULA

The making of the
Energy and Chemical Workers Union

Wayne Roberts

Between The Lines
1990

Published by Between The Lines
          394 Euclid Ave., #03
          Toronto, Ontario
          M6G 2S9
          Canada

Cover Design by Skip Hambling
Design by Skip Hambling
Typeset by Cheriton Graphics Systems Ltd.
Printed in Canada by the Alger Press

Between The Lines receives financial assistance from the
Canada Council, the Department of Communications,
and the Ontario Arts Council.

CANADIAN CATALOGUING IN PUBLICATION DATA

Roberts, Wayne, 1944-
          The making of the Energy & Chemical Workers Union

ISBN 0-921284-30-6 (bound) ISBN 0-921284-31-4 (pbk.)

1. Energy and Chemical Workers Union - History.
2. Trade-unions - Chemical workers - Canada -
History. 3. Energy industries - Canada - Employees
- History. I. Title.

HD6528.E472E637 1990   331.88'166'0971
                                        C90-095121-4

CONTENTS

PART I / COMBINING THE ELEMENTS

Section 1
Once Upon a Time It Was...1954

Section 2
A Long and Winding Road

Section 3
A Striking Difference

Section 4
Solidarity and Independence

PART II / APPLIED CHEMISTRY

Section 1
On the Job

Section 2
Bringing It All Back Home

PREFACE

The idea for this book came out of a landfill project that cost me the softest job I ever had. In 1984 I was at the Ontario labour ministry's Quality of Working Life Centre, paid to daydream about democracy and dignity at work. That's when I heard about the Energy and Chemical Workers, one of the few unions to go out on a limb for these ideals. I worked with their director, Neil Reimer, to write a call to arms for "bread and roses too," which went into the centre's magazine. The next thing I knew, a bundle of magazines was in the dump and I was looking for work. Reimer was about to retire as director of the Energy and Chemical Workers and asked if I'd write something about the history of the union and its forty thousand members. We figured a 20-page pamphlet would do just fine.

I ended up doing research for this "pamphlet" over four years while I was assistant to the president of the Ontario Public Service Employees Union, paid to put my rosy ideas into practice. The history-writing project became an excuse for stealing ideas from the ECW. I grilled ECW old-timers for ways they handled the nuts and bolts demands of running a union, how they dealt with internal conflict, set up bargaining strategies, worked out programs on new issues, started public campaigns. The research combined nitty gritty workaday details with a trip back to the future. From health, safety, and the environment to new technology to political action, there wasn't a new issue of the 1980s that ECW hadn't tackled 10 or 20 years earlier. Reimer and I agreed we'd have to expand the pamphlet to 80 pages to fit that stuff in.

The writing took two years, while I was a Queen's Park columnist for *NOW* magazine in Toronto, paid to peck away at groups that were copping out on big issues like poverty, the environment, and the future of Canada. I got tired of petty union squabbles and backroom deals, the bravado of make-believe campaigns, reruns of tired rhetoric, all symptoms of a movement hiding its fear of the future in bureaucratic busy-work. That made me see more in the ECW's history than a string of labour's past trials and glories. I saw a different union culture at work, one inspired by core values, open to fresh ideas, willing to listen, to learn, to reflect, to work differences through, driving hard to replace bitterness, grumbling, and apathy with education, power, and responsibility. I see that culture as the secret of the ECW's success and a training ground for a new political culture.

In the refining industries, where most ECW people work, the companies can pay well and can afford union wages. In fact, they've often paid union wages just to keep unions out. To survive in the energy and chemical industry, unions had to be pro-active: They had to talk about more than wages, they had to campaign for positive alternatives, to think big. The ECW had to take on big questions of national identity, the status of Quebec, and the relationship with the International headquarters — as well as more everyday questions about the daily grind on the shop floor.

The union's history, I found, is a laboratory of broken rules for success. It's also something of a showcase of what a union can do if it taps into unconventional workplace and community resources to fuel a made-in-Canada social movement. I have seen the past, and it worked. In the end, Reimer and I agreed, it would take a full-scale book to tease that culture out of the tangle of rushed events, knock-down-drag-out feuds, and the clash of strong-willed, colourful characters.

So, that's how I discovered the Energy and Chemical Workers Union. The rest of the book tries to explain how they discovered themselves.

AND THANKS...

Like most union projects, this book has been a collective effort.
To mark Neil Reimer's retirement in 1984, the Energy and Chemical Workers asked him to collect materials for a history of unionism in their industry. Reimer conducted interviews with 50 old comrades from across North America who had worked with him over his 40 years as a leader. The interviews were faultlessly transcribed by Jean McKinley.

Reimer, a compulsive packrat, also cleaned out his garage and attic and arranged for Allan Shandro, a University of Alberta librarian and historian, to catalogue 50 boxes of union records. The collection, now at the ECW headquarters in Edmonton is probably the most complete internal record of any union on the continent.

The union also funded my research, which included two hundred hours of interviews with Reimer and another hundred hours with friends and enemies whom Reimer felt he couldn't interview objectively. Eric Mills, Shea Hoffmitz, Laura Hollingsworth, Graham Murray, and Lori Stahlbrand helped me go through mountains of newspaper clippings and government documents to put the interviews and union records in context. Marc Zwelling donated his draft biography of Grant Notley, Reimer's close associate in the Alberta New Democratic Party. All of this material has been donated to the ECW collection.

The book centres on people and problems, not dates and places. Many of the people have passed away, but the problems haven't. I wanted this to be an open book, not a dated book, relevant for activists as well as historians. Activists need books that can be picked up and put down. So I tried to make each chapter and each section as self-contained as possible. To design a book that can be put on fast-forward, I relied on the imagination and talent of Skip Hambling, freelance designer. And Robert Clarke of Between The Lines burnt the midnight oil to edit out my cliches and misplaced pronouns and clarify my arguments.

The book doesn't provide reference notes, but scholars can easily check my sources. Shandro's catalogue offers an easy guide to past reports and letters, and McKinley has catalogued all the interviews. I have donated my research notes for each chapter, with complete bibliographic details, to the ECW collection.

This is a commissioned history but not an authorized or sanitized one. The union's executive board reviewed the text for accuracy but asked for no editorial changes. Needless to say, any errors of fact remain my responsibility. Bizarre opinions, malicious judgements, and silly flights of fancy are entirely my responsibility. My utmost respect and thanks go to the union leadership for allowing a known labour dissident to write a "warts and all" history of their union.

Wayne Roberts, Toronto
March 1990

PART

1

COMBINING THE ELEMENTS

Neil, can you

It was April Fool's Day, 1954, and it sounded like a prank.

Jack Knight, international president of the Oil Workers' International Union, was on the line from the headquarters in Denver. He wanted to talk to his youngest staff rep, Neil Reimer in Edmonton. "Neil, will you run Canada for me?" he asked.

Knight had just found out that the Canadian union was in shambles, run into the ground by internal bickering and mismanagement, going nowhere. He knew very little about Reimer, except that he was considered a young buck, a naive "stubble jumper from Saskatchewan," a little too prickly and keen. But Bill Ingram, Canadian representative on the international union executive, vouched for Reimer. They'd worked together at the Consumer's Co-operative Refinery in Regina — the world's first farmer-owned co-op refinery — and Ingram said Reimer was a workhorse who stayed clear of cliques and factions. That's all Knight wanted to hear.

"How long do I have to decide?" Reimer asked. "As long as we're on the phone," Knight answered.

An offer like that concentrates the mind.

run Canada ?

Once upon a time it was - - -

THE TELEGRAM   *Night*

2 MILES TO GO!

THOUSANDS LINE SHORE TO CHEER MARILYN HOME

THE TELEGRAM   *Night*

ONLY YARDS TO GO!

THOUSANDS CHEER MARILYN HOME -- WINNIE OUT

1954

When the trends and events of 1954 are put on fast-forward, the year shows up as the best and worst of times to start a union career.

Few years match it for sheer rush and confusion. Those who look back in anger or nostalgia misjudge its spirit. The year Marilyn Bell became the first person to swim across Lake Ontario was innocent, not smug, trend-setting, not conformist, charged with nervous energy, not laid back with complacency.

Rebels without a cause are box-office magic. Elvis the Pelvis makes his first records at Sun Studio. They're lining up to see Marlon Brando in *On the Waterfront* and *The Wild One*. They're jiving to "Shake, Rattle and Roll." Just a year earlier, "How Much is that Doggie in the Window?" had stolen their hearts.

It's the take-off point for life in the fast lane. Pizza had just become a craze. The first TV dinners and frozen baked goods are coming out of the oven. Consumers shop in the first indoor malls.

It's an age of miracles, and the high priests are inventors. In medicine, it's the polio vaccine and the first open-heart surgery. In industry, the first 30 business computers run off the IBM assembly line. Silicon transistors, which will make these huge computers obsolescent, have just been invented. The metal is still hot in the first oxygen furnaces that will revolutionize steel production. Oil and chemical plants set the pace for the latest and best in equipment. A brand-new $70-million Canadian Chemicals plant in Edmonton. Sherritt-Gordon's $19-million refinery in Fort Saskatchewan. A $4-million expansion of the Montreal Shell plant to give it the latest in internal coolers.

These are hard-driving times. A new line of gas-guzzling V-8s leads the pack in a record year for car sales. The government is paving the way with massive highway projects. Cape Breton gets its first road link to the mainland.

All that drive takes energy. From new hydro stations breaking ground for the St. Lawrence Seaway. From power lines stretched under the St. Lawrence to reach copper mines in Northern Quebec. From uranium just discovered at Elliot Lake. Experiments to tame nuclear energy as a power source have just begun and experts predict it will be "too cheap to meter." For the first time Canada produces more oil than it imports. Refineries have doubled their runs since 1947, when the legendary Leduc well washed Alberta in oil. The latest find in Pembina rivals all Texan reserves.

It's a time for positive thinking, not dangerous thoughts.

Liberal prime minister Louis St. Laurent, re-elected in 1953, is known as "Uncle Louis" and "chairman of the board." He's a comforting gray-haired grandfather figure who runs the nation's business with the help of technicians, not inventors. "They treat the national mind much as the officials of a trust company treat the mind of a rich widow whose funds they have been hired to manage," Hugh MacLennan writes of the government in *Maclean's*. "Ply her with accurate reports couched in a jargon she cannot possibly understand, but whatever you do, don't let her get too inquisitive about what goes on behind the doors."

"Such an unnaturally comfortable time could not last," Bruce Hutchinson, one of the country's leading pundits, recalls in his autobiography, *The Far Side of the Street*. The greatest new home comfort of them all, television, brings politicians into close range. It issues summons to the powerful and secretive, puts them under the spotlight, opens up political debate.

If politics don't look to the future, it's partly because no one wants to remember the past. James Gray, a leading oil-industry journalist, tries to raise fears about what's happening to oil but keeps bumping into a wall of dread. "For Albertans generally, the mere thought of going back to pre-Leduc economic conditions is enough to close all their receptors to suggestions that their country was being sold out from under them," he remembers in his autobiography, *Troublemaker*.

Fear of pricking the bubble on the economic boom is matched by the fright of upsetting the balance of terror in the cold war. In 1954 the cold war almost turns hot. Vietnam expelled its French colonizers. The U.S. president almost pulled the atomic trigger. McCarthyism still runs wild south of the border.

The threat of war isn't far away. Alberta will be "bomb alley" in the event of a U.S.-Soviet missile exchange. No home is secure without a fallout shelter. RCMP officer John Dowie scours movements of dissent to find Soviet agents. Peaceniks who lull people into a false sense of security are just as dangerous as Communists, he says. "People could be seen as a threat in many, many different ways," he says. The RCMP keeps a sharp eye on communications, oil, and transportation workers. The workers may defect with trade secrets or sabotage the economy with strikes, Dowie says.

The cold war has its civilian casualties. Oil companies cash in on the conformity and mistrust and pressure workers to sign an anti-Communist oath. As oil worker director, Neil Reimer will remind workers that their views are their own concern in a democracy, and they will refuse to sign. But the Canadian company Delhi Oil stands by its forms that ask about the country of birth of applicants' relatives, and about the fraternal societies and labour organizations they belong to. In these innocent days, workers are screened for beliefs, not drugs.

Human rights, civil rights, minority rights, and privacy don't count for much in 1954. Only Saskatchewan has a human rights code. Immigrants from Eastern and Southern Europe are lumped together as DPs, stuck in the worst jobs, expected to fit into a British country. Labour laws are weak, and they forbid unionization of civil servants. In Quebec, Premier Maurice Duplessis has just passed bills 19 and 20, outlawing unions that "tolerate" Communists or "threaten to strike public services." That gives him, according to labour researcher (and later senator) Eugene Forsey, "a complete veto" on union officers and staff. Even in Saskatchewan, an oasis of progressive policy, the government stonewalls every demand from power workers and threatens compulsory arbitration.

The days when workers had nothing to lose but their chains are over. Real wages are up 18 per cent from five years ago. The 48 - hour-week is on its way out. Half of the blue collar workers now enjoy a 40-hour-week, up from a third of them five years earlier. With money and leisure come the basic comforts. Two-thirds of homes now have flush toilets. Spending habits are changing. Personal expenditures on consumer products are double what they were in the roaring twenties, sparking massive expansion of the service sector and creating what a 1957 federal royal commission on economic growth calls "an acquisitive society."

It's not a year for bargaining dangerously or thinking dangerously. Unions no longer have to fight every inch of the way. There are only 162 strikes, involving less than one union member in twenty. Many leaders have made their peace with slow and steady progress. "The cold war has turned them into cats covering their own shit to show how clean and respectable they are," Reimer says.

Quebec is the exception. Unions there fight for their lives, are "at the very heart of the fight for progress and freedom" against the authoritarian regime of Duplessis, labour journalist Gérard Pelletier writes in his memoirs, *Years of Impatience*. The Quiet revolution of 1960, Pelletier says, was "the finish line," not the beginning of revolt.

Unions are less organized than employers. There are 111 separate unions, split by different shades of loyalty to craft, industry, or region, and linked to any of seven warring central labour bodies. The big split is between unions that organize workers on a craft basis — like the machinists and plumbers — and unions that organize all workers in an industry — like auto and steel workers. In 1954 the two camps are just starting to speak to one another, after 14 years of bitter infighting and raids.

The major craft unions go back to the 1870s and 1880s. The major industrial unions go back to the 1930s and 1940s. Whatever problems they have, they all enjoy a base and some momentum to carry them through the 1950s.

Not the Oil Workers International Union, spearheading the drive to organize energy and chemical workers. In this unlikely decade, it has to start something. In 1954 the union is in limbo, scrambling for a toe-

hold. The union and industry came too late for the gigantic wave of blue-collar industrial organization in the 1940s. They came too soon for the white-collar and public-sector unionism of the 1960s and 1970s. The workforce is labelled blue-collar, but work patterns are closer to science fiction than assembly lines. It will take sociologists a decade to coin a term for them — the new working class.

In 1954 the Oil Workers have about 2,000 Canadian members in 25 locals. They have just met for the first time as a Canadian council. It took 70 per cent of their budget just to cover travel costs. The union's future looks dim. It had struck a major Quebec employer in 1952 and got whipped. It had failed to organize an entire company chain in the refining industry. It had virtually no members in Alberta, the oil capital of the country. Its claim to represent gas and chemical workers wasn't recognized by the central labour organization in the country, the Canadian Congress of Labour. Its director, Alex McAuslane, had been a hero of the pioneer days of labour organizing but appeared to be a washed-up drunk in 1954. Its staff were at each other's throats. Its international organization lacked the resources to help out.

12

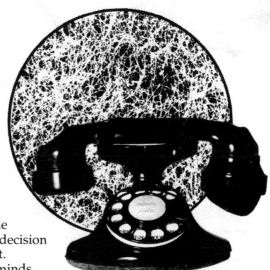

Knight is still on the line, waiting for Reimer's decision on this plum appointment.

Reimer was of two minds about his own future. He was a rebel, a maverick, a loner, an outsider, not exactly a joiner or a leader. But he also wanted to leave his mark on the world, to make a contribution, and that took organization. The job might mean a choice between being in the inner circle and listening to his inner voice.

Reimer came by his non-conformist personality honestly. In 1927, his Mennonite parents left the Soviet Union for Saskatchewan so they could keep their peace, follow their own conscience. They felt they owed Canada for that chance and raised their kids to pay off the debt by making some kind of social contribution. Making a contribution, but making it on his own terms, kept landing Reimer in trouble. In 1941 he had to quit university when he refused to practise killing during military training.

After nine years working for the Regina Co-Op Refinery, he went on the Oil Workers staff, in 1951, to try and prove that prairie boys knew more about building a union than the Canadian headquarters staff in Toronto. That attitude got him the booby prize, an assignment to Alberta where the union didn't have one member. Now, 3 years later, the union's international president wanted him to take on all of Canada.

Reimer's mind raced. His memory flashed to his Regina workmate, Walter Smith, an old Welsh coalminer who emigrated to Big Sky country to find scope for his imagination. Smith had pressed Reimer to go on union staff in the first place. "There aren't many people who have an opportunity to build something from scratch," Smith said. "If you do something, you'll know you did it. If you fail, nobody will know the difference."

Reimer took the dare. He told Knight he'd take the job.

"I got a fancy title, no members, and a mess on my hands," he says.

STANDARD PRACTICES

Whether oil tycoons are crude or refined, "roughnecks" or "establishment," they share a "mystical faith in the power of money to govern human affairs," Peter Newman states in his bestseller on Canada's rich and shameless, *The Acquisitors*.

Actually, their faith isn't that mystical. It's grounded in a century's experience of grasping power through monopoly.

In 1981 the government's combines investigator charged oil companies with "predatory pricing" that extra-billed Canadian consumers by $18.9 billion since 1958. There was no competition, just a few companies that closed ranks, fixed bids, fixed prices, and swapped supplies — all to "develop, protect and exploit their monopolistic situation," as the combines investigator showed.

The monopoly in the oil industry dates back to the 1880s, when John Rockefeller of Standard Oil, commonly known as a "robber baron," set the standard for rough trades against competitors and workers and insider trades with his own empire of subsidiaries. Collusion dates back to 1932, when six of the world's richest oilmen met at Achnacarry Castle in Scotland for a little grouse hunting. The oilmen aimed to keep cheap Middle East oil from driving prices down. Such cheap oil was "destructive rather than constructive competition," they agreed. The hunting partners pledged not to poach on each other's share of the world market.

In Canada the Big Six had bought out the industry lock, stock, and barrel by 1964. Standard hitched up with Imperial. Texaco bought out Regent and McColl Frontenac. Shell took over Canadian Oils and North Star. Gulf grabbed B.A., Highway Oil, and Royalite. Local figureheads gave a show of independence from foreign parents, but this was "landscaping to hide a mass grave of formerly Canadian-controlled companies," as James Freeman put it in *The Biggest Sellout in History*. Freeman's book, published in 1966 by the Oil Workers' union, was the first rallying cry against Americanization of Canada's economy.

Freeman quoted a 1962 oil analyst who admitted control "is now out of the hands of Canadians, except possibly at the political level." In fact, even the political loophole had been closed off in the

1950s, when government pipeline licenses gave the Big Six a lock on transport and the National Oil Policy carved up markets to suit parent company needs. Oil companies controlled all key government agencies "regulating" their industry, the 1981 combines investigation showed.

In the 1970s, government licenses, grants, and tax write-offs made political connections even more central to monopoly privileges. Gulf spent $1.3 million oiling Canadian political parties between 1960 and 1974, and Exxon spent $200,000 a year, Robert Engler shows in *The Brotherhood of Oil*. Marshall Cohen headed a Gulf holding company after presiding over the finance ministry that, according to the October, 17, 1975 *Globe and Mail*, had granted Gulf a billion dollars in tax savings.

THE WAY THE MOLECULE CRUMBLES

There are five ways to make more profit from chemicals and plastics, the *Financial Post* stated in a June 11, 1966, special issue on the industry. Companies can integrate forward, integrate backward, integrate sideways, build larger plants, and use advanced know-how. All five ways — but especially the first three — explain why conglomerates are the dominant form of monopoly in this industry.

The chemical industry doesn't lend itself to the same style of monopoly as oil and gas. Depending on the refining method, oil makes a range of similar products from propane to gasoline -- less than ten in all. Individual chemical companies make thousands of different products, from explosives to paints to aspirins, each with their own manufacturing method and sales market. Companies move up, down, and sideways with the molecules they split so they can control their input costs and rack up profits at each stage of production. "If chemistry as a science has made mankind less dependent upon a few naturally occurring substances," the 1957 Royal Commission on Economic Growth noted, "it has, at the same time, brought industries closer together."

Put less kindly, the chemical companies get the consumer coming and going. "If the seller controls the buyer," Alberta's 1968 inquiry into gas marketing noted, "it is easy to negotiate a price which is satisfactory to the seller." It was in the 1960s, according to John Blair's *The Control of Oil*, that oil companies went into petrochemicals. In 1964, for instance, oil companies spent $500 million to buy up Canadian fertilizer plants.

Aside from this industrial logic, chemical investors like to have portfolios — a few side-bets in other industries — to protect them from the ups and downs of consumer demand for any one set of products. Thus, Du Pont went into mining, and CIL went into real estate. The conglomerate nature of the industry came to public attention in 1975 when Power tried to buy out Argus Corporation, which owns the chemical giant Domtar. The power trip sparked the royal commission on corporate concentration.

While companies compound profits through complex integration, unions jostle for position, or jurisdiction, at the tail end of the process. Thus, George Eaton's 1968 study of chemical bargaining for the federal task force on labour relations counted 12 unions in the chemical industry and

19 in plastics fabrication. Melvin Rothbaum's 1962 study, *The Government of Oil, Chemical and Atomic Workers,* called this "competitive unionism." It defined the central problem of union organization in the industry.

Apart from suffering split jurisdiction, workers in the industry have multiple personalities. There is no common product, technology, or condition that binds chemical workers together. "To overcome this, I was always looking for common denominators like national bargaining, pay for knowledge, or osteoporosis, so the union could have an identity and joint campaigns," Neil Reimer says.

17

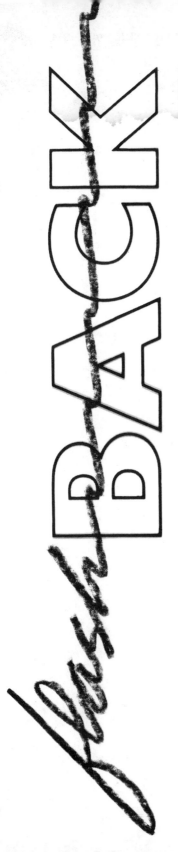

flash BACK

1935

CHEMICAL WARFARE

Organizing the energy and chemical industry was never a matter of A-B-C. Making sense of the early unions uses up most of the alphabet. AFL, CIO, CCL, NUOW, OWUC, UOWC, NUPW, just for starters. Then on to tongue twisters like UMWIU, UGCWIU, ICWU, and OWIU. They're all easier to name than explain. They're all easier to explain than justify. The titles represent a spelling-bee of bad organizational chemistry, personal power-tripping and chaos.

JOHN L. LEWIS

19

Reimer never was a man of letters. He saw the problem as too many unions for too few members, too much monopoly among employers and not enough among workers. No other industry was carved into as many fine slices as this one. No other union had to survive on the slim pickings left by older, more aggressive unions. For 30 years Reimer had to cope with the legacy of that labour civil war.

The family tree of chemical unions can be traced back to 1935. That's when John L. Lewis, the Winston Churchill of North America's unionized coalminers, marched across the floor of a U.S. labour convention, butted out his cigar, and flattened John Hutchinson of the Carpenters, the strongman of labour conservatism. The punch was felt across the continent. It put millions of workers in a fighting mood. Lewis became the personal symbol of the magic, hope, and determination of a new Congress of Industrial Organizations, formed right after the punch-out.

The CIO was a crusade to organize the unorganized. For immigrants, minorities, and unskilled factory workers who'd been scorned by employers and elitist craft unions alike, it was the great equalizer. The CIO had a simple principle: the power of wealth had to be matched by the power of numbers. It had a simple formula: one union for all workers in one industry.

The CIO scored direct hits in rubber, steel, auto, and meat plants across the United States. In other industries, such as chemicals and textiles, it shattered in all directions.

Lewis took a crack at factory organizing when he set up District 50 of the United Mineworkers of America to round up all the above-ground plants that used coal byproducts. District 50 was old before its time, set up just when kerosene lamps were crated off to antique shops. By the 1940s coal was no longer king of the energy and chemical industries. Oil and gas had beaten out coal, both as a fuel and as a source of artificial chemicals. To make matters worse, Lewis ran District 50 as a family business. He put his daughter in charge. The factory workers couldn't vote at Mineworker conventions.

In 1940, in a fit of rage, Lewis broke from the CIO and took District 50 with him after CIO leaders backed U.S. President Roosevelt. To retaliate, and to protect its turf, the CIO set up the United Gas, Coke and Chemical Workers Union. Its main business was raiding District 50, wooing its members back to the CIO. The new union was created from on high, but it allowed home-rule once plants were organized.

Then the old craft unions in the American Federation of Labor jumped on the industrial bandwagon. Until 1935 they had looked down on common factory workers. These factory workers, the craft unionists thought, had no skill, no class, no clout. By the 1940s the craft unions had realized they'd been too snobbish for their own good. The CIO got all the attention, all the new recruits. As the AFL lost standing, it rediscovered labour brotherhood. It sponsored its own industrial unions, including the International Chemical Workers, to welcome back its long-lost relatives.

In the United States the AFL ran the ICW as a "sweetheart" operation. AFL construction tradesmen building chemical factories traded influence to have the plant employers recognize the ICW voluntarily. That way, the employers cut off the CIO and got what they considered a more respectable union. The construction unions got first pick of plant tradesmen, and the ICW got the leftovers.

As a result of top-level feuds that had little to do with the needs of chemical workers, three unions — the International Chemical Workers, the Mineworkers, and Gas Coke — competed for members in the chemical industry after the 1940s. These random divisions undercut the principle behind the CIO's initial success: the solidarity of one union for one industry.

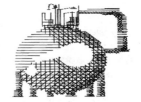

Chemical warfare was exported to Canada through the franchise system, known as international unionism, that each competing group sponsored. Each chemical union staked out claims for Canadian jurisdiction and set up shop near key border points, especially Sarnia, Ontario, in the chemical valley of Canada.

U.S. unions were able to establish subsidiaries because they had a ten-year lead on Canadians. They owed their head-start to the 1935 Wagner Act, which outlawed anti-union dirty tricks and made companies recognize unions with majority support. Canada didn't get its version of the Wagner Act — P.C. 1003 — until 1944, in the midst of the wartime economic boom. By then the law just recognized reality. Workers had already rushed to join unions in the same make-do way they'd played pickup hockey or sandlot baseball as kids, without much regard for the fine points of labour theory. They signed on with any U.S. team that sent in organizers and purused the fine print later.

The three unions, each the product of accidental tourism, battled it out for decades. Money was wasted, energy diverted, and ambition distorted on a scale unknown to unions in other industries. Leaders thought and schemed like hangers-on to U.S. coat-tails and big frogs in little ponds.

Such was the scene when Reimer became director of the Oil Workers in 1954, and things would stay much the same for another 20 years. Mistakes of the past dominated unionism in the industry until 1975, when rebels declared independence and moved towards one union for all Canadian energy and chemical workers.

STANDARD DEVIATION

Union organizers in the oil industry moved at a donkey's pace.

That's because they had to deal with a special breed of employer-sponsored "company unions" known as "joint industrial councils" or "donkey councils." Labour people held these councils up for contempt, but in the oil industry the councils spoiled workers to death. As a result, the Oil Workers International Union missed out on the mass organizing drives of the 1940s.

Company unionism was invented by John D. Rockefeller, patriarch of Standard Oil, the mightiest in the industry. Rockefeller always had more money than friends. Farmers and small businessmen called him a "robber baron."

In the ragtime days before World War I, Rockefeller symbolized the fear and loathing monopolists had for unions. He literally killed a union in 1914 when his company police machine-gunned a tent village of Colorado strikers and their families.

To counteract public outrage at this slaughter, Rockefeller turned to another hired gun, labour consultant and future prime minister of Canada, William Lyon Mackenzie King. King advised Rockefeller to dress the part of the philanthropist and offer workers an in-house council that settled troubles without resort to unions or strikes. This scheme gave Standard Oil the best of both worlds. It gave Rockefeller a new, caring image. It gave workers a feeling they counted. Both were part of the makeover of the Rockefeller reputation — "You can no longer just take from people. You have to make them want to give it to you," is how biographer Robert Scheer put it. But industrial councils gave workers none of the independence, clout, or resources that came with real unions.

The Standard plan had many imitators during the 1920s and early 1930s. But they only lasted in the oil industry, which had the money to volunteer the highest wages in town, to kill unions with kindness. In Canada, Standard's Imperial Oil was the first to set up works councils. Other oil companies followed suit. The result was a spotlessly union-free environment throughout the twenties and thirties. No local recorded in Harvey O'Connor's history of the Oil Workers International Union

1948

was started in those years. Even during the war, when company unions fell like ninepins in other industries, they held the line in oil. Amazingly, they remained a standard feature of the industry into the 1980s.

When the Oil Workers International Union started in Canada in 1948, it had three strikes against it. It was a runt union with a small industrial base to start with, because refinery workers were so few and far between. It was a rump union that could barely cover its backside in competition with company unions. And it was on the wrong end of the emerging petrochemical complex, without the resources or license to follow oil companies into their new growth area.

Oil drilling crew, 1916

THE ONE-EYED KING

The Oil Workers were the late bloomers of the industrial union movement. but their first Canadian director, Alex McAuslane, bloomed early. Equal parts scoundrel, romantic, and prophet, he was every bit a labour organizer of the heroic age, which peaked in the mid-1940s, just as oil workers came on the scene.

Born in Scotland in 1903, McAuslane first packed a union card when he was 13 and went down in the mines. He left Scotland for the merchant marine and was next heard of organizing embattled loggers and miners in British Columbia. Somewhere along the line he lost an eye. If you believed McAuslane, he lost it many times -- in the Irish Easter Rebellion, the Everest Massacre, the Winnipeg General Strike, all legendary battles of militant labour. Legend has it that whenever he addressed a union meeting he would pluck out his glass eye and baptize his audience in the blood of the cause. Before going to bed, he'd place his eye on his wallet. A good Scot, he liked to keep an eye on his money, he'd say.

McAuslane saw the country during World War II, as chief recruiting sergeant and trouble shooter for the Canadian Congress of Labour

(CCL), formed in 1940 to spearhead the organization of industrial unions. He stood at plant gates and mass meetings, a tall and stocky broad-shouldered man with a reddish complexion, booming out the message of a new land of brotherhood and plenty and collecting union cards. In the bars, he'd sing of the new day with a deep baritone that knew every labour folk song and light opera off by heart. Facing the music was a rich part of the comraderie he preached.

The CIO of his era didn't just turn on the concept of organization by industry rather than by craft. It turned on worker solidarity, without regard to skill, race, colour, creed, or sex. It cast out the beliefs often linked to the "American Separation of Labour," usually called "business unionism" in contrast to the "social unionism" of the CIO. The CIO was a crusade for social justice, not a machine for collective bargaining. It was a fighting movement that stood for community involvement and political action, "democracy in overalls."

But, like all good organizers, McAuslane worked himself out of a job. When he started, mass union memberships existed only in the minds of organizers. That's why CCL affiliates were called Steelworkers Organizing Committee, Autoworkers Organizing Committee, and so on. But once the CCL had organized workers it was supposed to transfer them to an appropriate union. That way, the Congress kept to the job it did best: lobbying government for all workers. And individual unions did what they did best: organizing and representing workers in one industry. The transfer process was logical, inevitable, and painful. The axis of power in the labour world shifted from a Canadian central labour body to sundry unions that were "international." The orbit of skills shifted from agitation to administration. Both shifts left warhorses like McAuslane without a home and without a cause. As steel, auto, and rubber unions began to fend for themselves, the Congress was left with the workers no union could be bothered with. Such were the oil workers. When World War II started, oil workers didn't even have an organizing committee. McAuslane organized a dozen oil locals on the fly. They were "tag-end" or "pickup" operations, with no geographic or chain-wide concentration. Around 1943, as the records of historian Harvey O'Connor show, a few oil locals claimed membership in umbrella organizations like the National Union of Oil Workers, the Oil Workers Union of Canada, the United Oil Workers of Canada, and the National Union of Petroleum Workers. But the CCL invented grandiose titles faster than it organized.

Organizing was a stop-and-go affair, mostly stop. In his book *Alberta Labour: A Heritage Untold*, Warren Caragata counted three attempts to start unions in Alberta's gasfields. Each time the local was dissolved almost as soon as it was chartered. The CCL's exasperated Western organizer finally pleaded for someone who knew something about the industry. His plea touched on the basic weakness of CCL efforts. No central labour body could match the expertise or staying power of an organizing drive by a specialized union.

Likewise, at the mammoth B.A. (later Gulf, then Petro-Canada) refinery in Clarkson, Ontario, local president Robert Kirk found CCL supervision "loose as a goose." There were no rules for running meetings, no newsletters, no co-ordinated communication or bargaining among separate locals, and no strike fund, he says.

These complaints about the CCL's helter-skelter style foreshadowed its post-war decline as a centre for ground-floor activism. The time came to shed that role and settle for a new role — one no less important but less inspiring, less direct, and in that respect, less powerful.

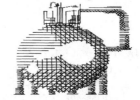

Until 1948 the Oil Workers International Union was international in name only. It had no patrons, like Lewis's Mineworkers, the AFL, or the CIO, to sponsor foreign expansion. Although the OWIU was a member of the CIO, the congress gave extra emphasis and support to Gas Coke. And Canada's refineries were too marginal to influence industry rates elsewhere, too borderline to be worth crossing the border. The OWIU was one of the last U.S. unions to enter Canada. It had turned down a CCL invitation in 1945 and finally agreed to check out the possibilities in 1947 after much pestering by McAuslane.

The OWIU was a rough and ready union raised in southern oilfields at the turn of the century. In 1935, Ben Shafer, its long-time secretary-treasurer, walked out with the CIO rebels after Lewis's dust-up with labour's old guard. In 1940, according to Ray Davidson's book *Challenging The Giants*, Jack Knight led a group of young Turks who won rank and file control of the OWIU. An executive board of working members were to hold the elected president accountable and stop staff from meddling in union politics. As a member-run union, the OWIU was more interested in service than in expansion and empire building.

In 1947 the OWIU sent its vice-president Sarge Kinstley out to look Canada over. Kinstley, a big tall Texan who came honestly to a grand scale of thinking, teamed up with McAuslane for a cook's tour of the country. They were a perfect team. Kinstley knew nothing about Canada and McAuslane knew nothing about the oil industry. "I think he and Sarge were two of a kind," says Charlie Armin, a retired OWIU regional director and friend to them both. "You take two guys that had a rosy outlook towards the future of labour, and they saw all of this imaginary or real potential some place. They never had anything happen in their lives that they didn't dramatize."

In 1948 McAuslane rolled out his Scotch "R"s at the Oil Workers' convention in Toledo and swept them off their feet. The U.S. unionists had never heard such a spellbinder. "Even his cursing was a colourful litany," says Ray Davidson, then a wide-eyed staffer at his first convention.

Both Kinstley and McAuslane laid out prophesies about how they'd get some 10,500 new members in Canada, mostly from the Northern Albertan tar sands, the next Texas of the North American oil scene, McAuslane's Promised Land. Neither Kintsley nor McAuslane got around to taking a bush plane to the tar sands. They preferred flights of fancy. Some 20 years later the area was opened to large-scale production. Some 40 years later the union enlisted a thousand-member local.

In 1948 the Oil Workers named McAuslane Canadian director and pledged seven organizers to put the new district on its feet. That was a eal boost. On their own, the fifteen-hundred Canadians in CCL locals couldn't afford one staffer. The OWIU asked for little in return. It even agreed to lower dues from Canadian members. But the Canadian district wasn't a U.S. subsidiary. The first locals were bootstrap operations, formed on Canadian initiative. They may have joined an international union but they weren't organized by one. And the international they joined was down at the heels. Only decades later could that be seen as a blessing in disguise.

After taking Canada under their wing, the Americans in OWIU went into a tailspin — the first victims of the post-war crackdown on labour. The Taft-Hartley bill of 1948 domesticated U.S. unions by requiring all labour leaders to swear a non-Communist oath. This government interference in what unionists thought "made us second-class citizens," says Charlie Armin, but "while it was bad for our soul, it meant nothing to anybody."

The Oil Workers launched a 1948 strike in California, then a union stronghold, before discovering other key clauses in Taft-Hartley. Supervisors could be ordered to work during strikes. Mass picketing was outlawed. Strikers weren't guaranteed their jobs back. During the strike, Armin says, company supervisors kept the plants running while police "handed out injunctions like they were pamphlets." The union was sued for millions in lost business. Strikers raised the white flag, but less than half of them were taken back. California membership dropped from 24,000 to 9,000.

The OWIU was flat on its back and dead broke. It cancelled its 1949 convention. It cancelled staff expansion in Canada until 1951. Canadian local delegates didn't meet each other, or U.S. leaders, until 1950, in a convention that turned out to be a riot of slander and abuse. With nowhere to turn, members turned on themselves. Knight quit after one chorus of insults, and the convention recessed until he changed his mind.

The eight Canadian delegates, meeting for the first time at this 1950 convention, were at loose ends. Reimer and Robert Kirk were both nominated for the Canadian spot on the executive board. They were both defeated by Alex Macleod from British Columbia, who was not even a convention delegate. MacLeod won a post-convention membership referendum as a write-in candidate. He thought that made him McAuslane's boss, which got their relationship off to a shaky start.

The OWIU connection, however, beefed up the quality of union service. B.A. in Clarkson, Ontario, may have suspected that. Kirk says B.A. shut the plant down for a few weeks to drop a hint of the trouble workers could expect if they joined the union. The threat didn't work, and the Clarkson local was the first chartered by the OWIU in Canada. With affiliation came such basics as a constitution, local bylaws, and a union newspaper, the *International Oil Worker*, with at least a page of Canadian coverage. The locals were in business.

The new union became a wage leader. A 1950 contract put the workers of Montreal's McColl-Frontenac (later Texaco, then Imperial) ahead of the whole province, according to an *International Oil Worker* report on October 9. A one-month strike against Shell at their Burnaby, B.C. plant in 1950 broke the Imperial Oil wage barrier, and shamed Imperial into recalling its company union for an 11 cent an hour raise, a whopping sum in those days.

The Canadian district, however, was a mess. McAuslane had lost his magic as an organizer. One time, according to Reimer, McAuslane started off in Vancouver and said, "Well, I guess I'd better go and organize this union." He scheduled stops in Kamloops, Calgary, and Edmonton, and thought that by the time he hit the Atlantic region there'd be an oil workers union. "That's how he viewed it because of his experience," Reimer says. "He came from the generation of `I came, I spoke, I conquered.'"

Time passed McAuslane by, says Ken Bryden, who was deputy minister of labour in Saskatchewan when he met the union director in the late 1940s. In the glory days, Bryden says, "You just called a meeting and whipped them up and they were ready for the new Jerusalem and singing `Solidarity Forever.' Organizing in those days wasn't the slogging business of going to one person after another, and trying to sell them on the union."

"He never quite got a handle" on his new union or the new age, Kirk says. "He was from the rough and ready era of the coal mines and steel mills where he'd organize a massive assault of bodies. You couldn't do that with oil workers. You had to be smooth. You had to base your case on evidence."

McAuslane had no feel for the mentality of oil workers. He recruited staff from outside the industry. That rubbed the wrong way, Reimer says. "The body rejects foreign parts, and so do workers, especially in an industry as technical and unusual as the petrochemical industry." Reimer, hired in 1951, was the first industry insider on staff.

Amazingly, McAuslane barely knew how to negotiate a collective agreement. He couldn't settle down long enough to write the briefs then required by government conciliators. Landon Ladd, another legendary figure from that era who knew McAuslane well, says agreements weren't important for their generation. The union leaders put little faith in a contract with the devil. At any rate, they felt, there wasn't a problem that couldn't be fixed by a wildcat strike. Except that after 1948, wildcat strikes were illegal everywhere but Saskatchewan.

Diplomacy and the work that went into building a relationship with management were not McAuslane's strong suits. He broke the ice with one company heavyweight, Blenner Hassit, by asking politely how he came by his name. It turned out that Hassit's great-grandfather Armin rode a greased pig in a barnyard rodeo. The crowds chanted "Armin has it," and the name stuck. "Well, I'll be," McAuslane said. "I've never met a genuine pigfucker before."

Intrigue was McAuslane's passion, which Reimer once turned to good effect. For their first contract talks, managers at Edmonton's Celanese branch-plant picked a starchamber crew of negotiators from the southern states. Reimer figured they'd been tipped off about McAuslane's foul-mouthed style, and that they were on orders to blow him away. Reimer convinced McAuslane to trick the company by playing sweet. Two days later, Reimer found McAuslane gazing at his naked body in a mirror, mocking his own sweet reasonableness, repeating "I can't recognize myself anymore. Is that me?" The Celanese international vice-president called Reimer to praise McAuslane, and asked how he got his terrible reputation. "Oh, managers like to exaggerate a little bit now and then," Reimer told him.

McAuslane was the victim of a 1950s-style generation gap. The heroic agitator who organized at breakneck speed had the wrong stuff for the twists and turns of a more complex, less dramatic decade. Industrial relations demanded a new set of union skills. Those with one eye were no longer kings. Modern labour leaders had to be strategists, able to link organizing with collective bargaining, to build coalitions that withstood hostile government and public opinion, to administer an organization with more members than most companies had employees, to shift focus from agitation to problem solving. The decade brought out all of McAuslane's weaknesses and few of his strengths. "When it came to the nitty gritty of administration, he was completely hopeless," Ken Bryden says.

Unable to find a vent for his explosive energy, McAuslane's inner time-bomb imploded on the union and his personal life. Being a rascal made him a local hero in the 1940s when it took a daredevil to defy the boss and daring-do to break the establishment's rules. Being a rascal in the 1950s turned a union outside-in. Infighting and jockeying for position became a way of life in the Oil Workers. Two organizers were fired for sloppy bookkeeping, accused of dipping into union strike funds. They counter-attacked with tales of nursemaiding "McQuisling" during week-long benders that turned him into a primitive brute. One staffer, according to Kirk, kept McAuslane supplied in booze with a view to pushing him over his self-destructive edge and replacing him as director.

Later, when McAuslane was a staff rep reporting to Reimer, he missed the grievance hearing of a local president, a sole-support father of three who was subsequently fired. Reimer sent McAuslane a note, saying he had to consider whether McAuslane was worthy to represent working people. Something in the note hit home. McAuslane took the cure, nursed back to health by Reimer's wife, Merry. There are reports that McAuslane kept Reimer's note with him, and opened it whenever he felt himself weakening. Though he became one of the union's ace troubleshooters, McAuslane never shook his obsession for organizational intrigue. He took early retirement, an offer he couldn't refuse, in 1967. In his sunset years he worked at a porn shop in Toronto.

McAuslane will have a place in the rogues' gallery of some future Labour Hall of Fame. But for the OWIU and its successors, he's more than a skeleton in the closet. He's a link to their CIO past.

The cold war and bargaining pressures of the 1950s led many union leaders to lock their CIO heritage in a time vault for safe keeping. The OWIU didn't have that luxury. As the last and weakest of the CIO unions, it had to maintain the CIO rallying cry to "organize the unorganized" throughout the 1950s. It was the only way to survive. As the last and weakest of the CIO unions, the Oil Workers turned to the youngest and least experienced union leader of the CIO generation to replace McAuslane. When Reimer retired in the 1980s, he'd already celebrated his tenth anniversary as the last labour leader in the country to trace his formative years to the CIO era.

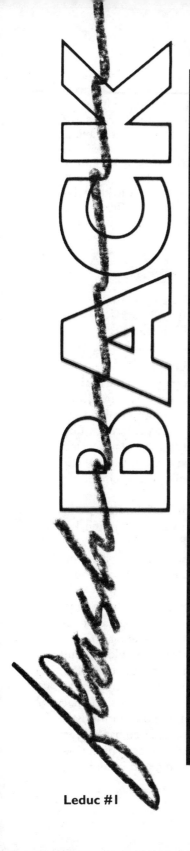

1951

LAND OF THE SECOND CHANCE

Merry Reimer started life in Edmonton cutting the air with a knife.

On her first day in town, in August 1951, the neighbourhood wives threw a welcoming party in the best tradition of western hospitality. And what does your husband do, Merry? That wasn't a trick question where she grew up in Saskatchewan. He's on staff with the Oil Workers International Union, she bragged. Dead silence. The welcome signs went down. So did the property values. On Sunday the minister stared straight at the Reimers when he blasted unions as irresponsible agents of discontent.

In 1951 Alberta was the death valley of Canadian unionism. That's why Reimer got sent there. "We didn't have a single member in Alberta," he says. "There'd been five reps in before me, and they all quit in exasperation. This was the graveyard."

McAuslane thought the assignment fit the crime. The year before, he had dropped in at the first meeting of the union's western

Leduc #1

33

council and got an earful from an upstart, with all of one year's experience as local president of the Co-op Refinery in Regina. Reimer told McAuslane to keep his nose out of local business and put his nose to the grindstone of organizing the industry giants. Local strikes are useless as long as multi-plant companies can transfer production elsewhere in the chain, Reimer lectured. Until the union cracked the plants in Alberta, existing locals would remain the ragtag and bobtail of the last decade's organizing. McAuslane was impressed enough to put Reimer on staff. But assigning him to Alberta was a nice way to keep the long-distance feeling between them, and to enjoy watching Reimer eat the dust of his own words in the drylands. McAuslane didn't visit Alberta until 1954.

In 1947 Alberta finally lived up to its reputation as the land of the second chance. After 20 years of drilling dry holes, an Imperial prospector nicknamed "Dryhole" struck paydirt with a wildcat drill at Leduc, 20 miles from Edmonton. Five hundred well-wishers came out in February to douse champagne on the orange flame and dense black smoke and kiss the era of one-crop farming and low-grade coal-mining goodbye. "It seemed as though a miniature atomic bomb had been dropped as the smoke formed a mushroom-like cloud and beat its way upward with a roar," the *Edmonton Journal* reported. The strike blasted Alberta into a kingdom come of new wealth and power and transformed its economic, social, political, and emotional landscape.

Oil and chemical companies rushed to set up new plants close to the oil patch. It was a hey-day for brokers, prospectors, and roughnecks, the first big break for up and coming executives.

In his autobiography, *Troublemaker,* James Gray recalls the "steadily increasing influx of a special variety of the species homo Americanus — eager-beaver young junior executives on the way to the top." They had militarist backgrounds, and they had the "awesome herd instinct of Americans." They set the tone for getting down to business, and getting rid of anyone who stood in their way.

The Social Credit government pandered to them. Oil blotted out the memory of the party's Depression roots and its promise to break the stranglehold of Eastern bankers by turning money over to the poor. Oil replaced "the people's money" as the cure-all for economic ills. All that remained of Social Credit's legacy was rage. Unions and Communists, to the extent the government made any distinction, replaced bankers as the bogeymen standing in the way of progress. Labour legislation was stacked against unions, which made both organizing and bargaining difficult.

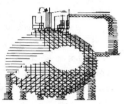

Roy Jamha, a leader of Edmonton's labour council and its biggest union, the meatpackers, offered to show Reimer around. Jamha soon felt like an outsider in his own town.

Jamha was used to having the run of the mill in the ramshackle packing plants. He had been able to call a wildcat strike or a sit-down at the drop of a grievance. Now he was overcome by the forbidding steel and concrete of petrochemical architecture. All doors were closed to him. "The plants all dealt with very secretive, volatile, and dangerous commodities, so security was extreme," he says. "You couldn't even get on the property, and you had to leaflet from the outside."

He didn't recognize any names or faces in boomtown. "Everyone was on rotating shifts, and the workers floated from one plant to the next," he says. "It was the easiest way to advance yourself to a higher position in an expanding industry. Plus, as a prairie kid, I didn't even know the names of half the things they were producing."

Reimer did know the industry, but he too had big problems adjusting. Fresh from the open and engaged politics of Saskatchewan, he couldn't cope with how closed Alberta was. People held their ideas close to their chest. They treated political arguments as akin to bad manners. Talk about unions had to be whispered.

He tried to explain the situation to McAuslane, who was still clinging to a big bang theory of labour organization. "People and government officials are amused when you tell them that you are here to organize oil workers," Reimer wrote in 1952. "The industry is new and men working together do not know who can be trusted and for that matter even his fellow worker's name. We first of all have to `create' a desire to organize and then put on the drive.

"Maybe a more experienced organizer would be able to take a few shortcuts, but I'll be damned if I can see them. Furthermore, as far as I am concerned I am unable to agree that we should design our organizing drives in such a manner so that they will suit these weekly report forms. We will just have to have patience."

Beginners' luck helped in two small plants. Reimer was invited to tour the boiler room of a bakery. The workers there shuffled when he asked why they hadn't joined the AFL's Bakery and Confectionary Union, which represented the rest of the plant. They were all ex-cons, and the AFL wouldn't have them. Reimer signed them up.

Reimer got his next 30 members at Liquid Carbonic. That got him a delegate's badge to the Industrial Federation of Labour, the coalition of CIO unions in Alberta. On his first night there everyone courted him. It turned out that he had the tie-breaking vote that decided whether the Communist Party or the democratic-socialist CCF would lead the federation. He'd already been lined up by Jamha to back the CCFers.

That was the only place he could cast a deciding vote. None of Alberta's 11 refineries were organized.

Reimer started at B.A. and quickly signed up almost two-thirds of the workforce there, enough cards to win automatic union recognition. Then the government ruled that the oil industry was too new and important to certify a union without an election. This meant trouble. In the United States elections are used to decide if a union has majority support -- it seems more open and democratic than collecting cards on the sly. But the procedure allows an employer to campaign, to create an atmosphere of fear that hampers workers' rights to make a choice on their own. It also gives the union drive a deadline that prevents the union from applying when it's at peak strength. During the B.A. campaign, the company sent officials to note the names of those who attended union rallies. On voting day, the union lost by ten votes.

The hurt of this first loss stayed with Reimer for life. He cherished it, developing a theory that it's good to lose early. "People advocating change get more knocks than successes. You need to learn early that workers will let you down," he says. "I've seen people have their first kick in the ass late in life, and that's the end of them. I got to appreciate that it's not really a loss if you persuade more people to go a certain direction in society, to make a commitment, which is really what the fight is all about."

Reimer owned up to his own mistakes. He'd connected with some old-time miners in the plant who did all the inside organizing. Their style was too brash and harsh for an above-ground operation. Oil workers didn't take kindly to orders from a union. As it turned out, the union supporters just up and quit B.A. after losing the vote. "It was almost a strike," says Reimer. "The company couldn't run its plant. Powerhouse workers like Pat Gariepy relocated at CIL and Canadian Chemical and helped us win both."

Reimer became more certain that Alberta was where an energy and chemical union had to be based, even if it needed to be kickstarted by the national operation. As soon as he was appointed director, he transferred the union's Canadian headquarters to Edmonton.

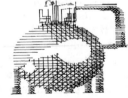

Almost half a century later, union members still debate the location of the headquarters. Reg Basken, who became president in 1984, admits Edmonton is removed from some of the action. But he says Alberta offers the vantage point needed by a far-sighted union. "Our union has to view energy from the producer standpoint, the researcher standpoint, the consumer standpoint, the worker standpoint and the manufacturing standpoint," he says. "Five points that in other provinces are only one or two."

In 1954 Reimer just wanted to go where no union had gone before. In Alberta he saw room for a late starter to pick up speed and come from behind. "Other unions weren't even aware the West existed," he says. "So I saw a chance to organize without competition, to get a base." Even a small union could make its mark on virgin territory, get its people on executive bodies of the labour movement as representatives of a region. The union got a leg up on the competing unions that were lost in the crowd out East, dwarfed by mass memberships in auto and steel. "I got on the inside of the labour movement, whereas all the other unions in our industry were on the outside, even though they had a headstart," Reimer says.

There were deeply personal reasons too. Raised as a Mennonite, Reimer was attracted to the lone west. "It's important for me to be at peace with myself," he says, "and I can't buy somebody else's thinking to do that. I've always had to be alone for a certain period to catch up with myself." He needed to be far from from the place-seekers, palace guard, and backstabbers who thrive in the rarified atmosphere of big centralized headquarters. "I can't tolerate headquarteritis. So we likely built the skinniest union in the world in terms of headquarters." And so he led his union into the wilderness.

The move west gave the union the space it needed to be different and the place to be with doers and thinkers instead of movers and shakers. It also gave the union an ear for country and western sounds seldom heard in the tone-deaf metropolises, and a feel for the hinterlands that create Canada's wealth but rarely share in the harvest.

The quality that set prairie towns apart, writes western editor James Lamb in his autobiography *Press Gang*, "was the intensity of their communal spirit. Mostly this was a reflection of their harsh environment, for isolation and severe weather engender a sort of closing of the ranks, and an all-in-the-same-boat togetherness." Breathing this in at union headquarters allowed the spirit of tight communities to counterpoint the norms of the ultimate plastic industry. It marked the quality of personal relations in the union and formed its political and bargaining agenda. Being based in the west gave the frontier a chance to civilize the centre.

The history of the union is a history of breaking the mould, of non-conformity, of oddballs. That tradition was solidified by the decision to stake the union's future in Alberta.

STRONGMAN

Maurice Duplessis was "the chief," the strongman, of Quebec. His main selling point to foreign investors was that he ran a distinct society that was meek.

In 1954 Duplessis blew his top when Herbert Lank, head of Du Pont, started to complain about a new tax. "In Quebec, we have two ways of attracting industries and keeping them here," he said, according to Gérard Pelletier's *Years of Impatience.* "The first way: natural resources," Duplessis said. "I don't sell them to you, I don't rent them to you, I give them to you." Then he said, "There's a second way: manpower. Quebec's manpower costs you less and gives you less trouble than any workforce in Canada, not to mention the United States. And my government, the Union Nationale, takes care of that for you. And you have the gall to come crying to me, you, the bosses!"

Duplessis made the same pitch to the Canadian Manufacturers Association, Conrad Black reports in his massive biography. "Quebec is the only place in North America where one can be sure that there will never be any communists, leftists or other radicals," Duplessis told them. "Our working class in Quebec is great, numerous, submissive to authority," he told one labour group.

Intellectuals such as Pierre Trudeau rebelled against that submission, saw it as the mirror of a closed, nationalist, church-ridden society. That view appealed to many English Canadians, who liked to blame Quebec's "backwardness" on the province's own narrow nationalism and conservative religion.

But Duplessis' policies were a far cry from nationalist or church doctrines. In *Quebec: Social Change and Political Crisis*, Dale Posgate and Ken McRoberts show that Duplessis' economic strategy followed "the laissez-faire economic liberalism of the business circles," not clerical nationalism.

A workforce that was "submissive to authority" was the key to the acceptance of economic liberalism. That's why unions that believed in workers' rights

1952

were such a threat to the regime. That's why Duplessis put such a high priority on forceful repression of strikes.

Duplessis championed Quebec autonomy. That was his Bible, René Lévesque, a leader of the last of the 1950s strikes, the 1959 strike against CBC, wrote in his memoirs. Autonomy was "the Maginot Line behind which nothing was supposed to change," Lévesque states. "Freedom? Never in your life, for however small the dose it could give people ideas."

From that perspective, those who walked the picket lines during the 1950s were the freedom fighters for a new Quebec identity.

▼ PART I
▼ SECTION I / FLASHBACKS
▼ CHAPTER 4

THEIR COUNTRY WAS WINTER

There's something odd about 550 people walking around in the middle of summer, looking forward to walking around in the shivering cold. On July 14, 1952, pickets surrounded the Canadian Copper refinery in east-end Montreal. They joked about icicles. They knew that their country was winter, that the company would wait out the strike 'till its pipes froze over. Only then would it start to deal.

In 1952 Quebec was still in the ice-age of labour relations. Premier Duplessis was the iceman. His Union Nationale party was re-elected with a heavy majority on July 16, two days after the strike at Canadian Copper began. By year's end Duplessis was using police-state methods to defeat three Quebec strikes that accounted for 75 per cent of all time lost to labour disputes across the country. Canadian Copper was one of them.

It's said that the famous Asbestos strike of 1949, led by a once-pious Catholic union and backed by a group of intellectuals including

OWILU Montreal area members welcome officers from international headquarters.

Pierre Trudeau, was a turning point in Quebec social history. By 1952 it was clear that the asbestos strike was a turning point that didn't turn. The ancien regime wasn't to be toppled by one strike. It became equally clear in 1952 that there was no turning back. Labour militancy wasn't just a passing trend after the war. It became the cutting edge of a new sense of Quebec.

Union militancy was partly a defensive response to Duplessis' labour regime, characterized by industrial relations specialist Carla Lipsig-Mumme as "coercive integration larded with simple and devastating repression." At the same time, the militancy laid bare the wide range of health, social, political, and cultural disorders that festered in the workplaces of old Quebec. It anticipated the aggressive and wide-ranging demands that later came to the fore during the Quiet Revolution of the 1960s.

If any one company symbolized the new-age technology that combined with the almost medieval labour relations in 1950s Quebec, it was Noranda. Noranda owned Gaspé Copper, site of the notorious Murdochville strike of 1957, which was fought over union recognition and marked by pitched shoot-outs between police and workers. The company ran the regions of Rouyn-Noranda and Abitibi and established what former Premier René Lévesque called a "Rhodesian climate." It also owned Canadian Copper Refiners, the world's second-largest copper refinery, running it in partnership with the U.S. multinational corporation, Phelps Dodge. Noranda sold the copper ores it mined to Canadian Copper, which in turn sold copper to its next-door plant, Canada Wire and Cable, also owned by Noranda.

Strikers at Canadian Copper, represented by Oil Workers International Local 636, couldn't match that power. But they went the distance and put in sharp relief the new forces that were transforming Quebec.

On July 12, 1952, copper workers met at their union hall. They'd been on edge for a year, ever since 85 per cent of them voted to have the Oil Workers speak for them. All that time, the company refused to recognize or bargain with the union. Local works manager H.S. McKnight, employed by the U.S.-based Phelps Dodge, refused to deal with a union that was "not Canadian and is controlled by outsiders."

The day before, the company had sent all workers a letter telling them that a court had quashed the Labour Board's ruling certifying the union. "Each employee should clearly understand that a strike at this time is entirely illegal," the letter advised. Then the company fired all local union leaders. The workers were in a bind. If they struck, they would be breaking the law. If they didn't, their union would be broken. A large majority voted to strike.

"Our union is leading the fight against the attempts of the bosses to smash trade unionism through court procedures," McAuslane wrote International headquarters in Denver. "The fight is important not only to the CIO-CCL but will be a determining factor for all other trade union centres operating here."

McAuslane's appeal wasn't just rhetoric. In Quebec a strike for union recognition was a strike against the province's labour legislation. It was so complicated, with so many compulsory waiting periods, that anti-union companies were able to hide behind delays to bust a union. Canada Copper's year-long suit against the Labour Board was a case in point; the fact that the company won was simply icing on the cake. This was labour law writ backwards, with a bias against strikes, not for recognition or equality.

But making the strike a test case of Quebec labour law made it a test of strength with the government. It hit the sore point of Premier Duplessis' strategy for political survival. Labour laws typified Duplessis' system for maximizing the entry points for government intervention — thus providing him with options for political favoritism — while minimizing the points of government regulation — thus providing corporations with maximum freedom of action.

The company lost face in the first week of the strike. Office workers walked out when they learned that they'd be kept captives in the plant for the duration. Morale on the picket line stayed high for four months. The plant was down. There were no problems with police. It was just a matter of waiting for the icicles to appear.

Walking the line gave workers time to become friends. They compared doubts and fears. That's how they learned about the effects of the acids and arsenic used to tear copper from the ore. About how common it was to have three pairs of pants a week torn to shreds by the acid. Men who had worked at the plant for five years had no teeth. They were all ashamed of their body odour. Sometimes they were asked to move away at shows. "You could smell a copper worker before you see him," a striker told one journalist. Worse, they themselves had to breathe the foul air in the plant every day. What was happening to their lungs?

These became the new issues of the strike. On August 7 union staffer Larry Packwood told a government official, "The demands are mainly to improve working conditions which are intolerable." Health and safety topped the list.

On November 15 the company tired of the waiting game. It gave strikers an ultimatum to come back or get out. On November 19 the company hired strikebreakers, mostly new immigrants — DPs, the strikers called them. They were picked up in ten yellow and orange buses that were rented for a thousand dollars a day and escorted to the plant as a convoy. Montreal police rode in front and armed guards from the Barnes and Atlas detective agencies rode shotgun. Strikers who tried to block the plant entrance were roughed up by police, then charged and fined.

On December 1, strikers demonstrated at Quebec City for stronger union recognition laws. On January 8, 1953, dynamite shattered a power pole just outside the company fence. The union denied involvement. Many of the strikers had been commandos during the war. They were insulted that anyone would accuse them of such shoddy work; the pole was only shattered, and the power flow was uninterrupted. Strikers said the bombing was a company ruse to justify police harassment of strikers.

Provincial police launched an investigation. They picked up Jean Paul Fournier on January 28 at 2:30 in the morning, on his way home from a double date. "We need you," police said, and took him in for questioning. They booked him under an assumed name so friends couldn't trace him. They worked him over all night, hitting him in the gut with nightsticks, slapping his face, and whacking his wrists with blackjacks. In the morning Fournier signed a form promising not to take action against the police for anything that had happened. He was released at noon.

Two other strikers got the same treatment. Aurele Mainville, a former commando, was held for 41 hours, rammed repeatedly in the gut with nightsticks, and released after signing a statement accusing union staff rep Larry Packwood of setting off the bomb.

That night, January 29, Packwood left the union hall at midnight, was picked up by police, and driven to St. Jerome. For four hours he leaned against a wall, balanced on his fingertips. He denied any connection to the dynamiting. At another station, striker Vincent Fitzpatrick was beaten with a hose and blackjack. He denied any involvement or knowledge.

Police charged no one and arrested no one. They conducted all the raids at night and under cover of intellectual darkness. Although the strikers swore out affidavits, and although their affidavits were never challenged, the English press in Quebec filed no reports. The French press provided brief coverage.

In Ottawa the *Citizen* carried a major investigative feature. An editorial denounced the "fascist-like conduct by the provincial police" and called Quebec a "police state." The Canadian Congress of Labour highlighted the case of Canadian Copper workers when it set up a defence fund to protect the rights of Quebec unions.

But it was too late to save the strike. The minister of labour, Antonio Barrette, pleaded with the company to "meet half way" by rehiring strikers and starting fresh negotiations. "Our plant resumed operations about two months ago," the company replied on January 9. "Former employees desiring to apply for work may do so in the regular way."

On March 11, McAuslane conceded defeat. The strikers are valiant, but the struggle is lost, he wrote headquarters. Strike pay was extended one week. In July the new company workforce voted out the union.

The defeat left a bitter after-taste. McAuslane accused two staff reps of pilfering donations sent to help the strikers. The charge was investigated by the international's executive board, which condemned the local's bookkeeping methods but upheld its honesty. Both reps had gone into hock for the strike.

Whether they saw union recognition or health and safety as their issue, the Canadian Copper strikers sensed they were fighting for a principle rather than a victory. "We fight for decency or we die," copper worker Henri Mathieu told a reporter. Many of the men were veterans. Local president Walter Reid had been wounded twice during the war. Henri Leduc, the recording secretary, and James Herrington, the picket line co-ordinator, had fought overseas. They thought they had won a war against police dictatorship in 1945. In 1953, they lost one.

Herrington's family was evicted when he fell behind on his mortgage. Their belongings were thrown on the street. He intended to demand police protection of his property. "That will be a laugh — getting police protection instead of persecution," he told the *International Oil Worker*.

Quebec has a long history of police repression of strikes. Textile, garment, and mine workers have all taken their knocks. This strike represented something new. It took place in the hub of Montreal's zone for heavy industry. It highlighted a fundamental shift in Quebec's economy.

As early as the 1940s, Quebec's industrial base had moved away from labour-intensive jobs in outmoded textile and clothing factories. In 1943 chemicals provided more jobs than any other sector. The trend to high-tech industries gathered speed during the 1950s.

This new economy created a base for aggressive organizing. It provided new choices, something other than the lesser evil of company bankruptcy or low wages. It provided more informed choices because Quebec workers learned about what similar companies were paying elsewhere. That's one reason why Quebec workers turned to international unions. Unions like the Oil Workers made a point of emphasizing the discrimination practised against Quebec workers. The November 1954 issue of the *International Oil Worker* publicized a study that showed how Quebec workers earned 15 cents an hour less than Ontario workers, and worked two hours more a week. Internationals were also a vehicle for industry-wide bargaining and uniform rates.

This was a decade and environment where Quebeckers could learn, as Sheilagh Hodgins Milner and Henry Milner put it in *The Decolonization of Quebec*, that poverty was not the destiny of their race but the product of power relations. In 1960, Antonio Barrette, the labour minister who oversaw the destruction of the Canada Copper strike, ran to replace Duplessis as premier. He lost. The Quiet Revolution began.

A long and winding road

HARDENING OF THE CATEGORIES

Jurisdiction is about the simple matter of turf, or market share — which union gets to organize what workers. But deciding jurisdiction is the greatest brainteaser of Canadian union history. That's because decisions about jurisdiction must weigh a host of conflicting priorities: solidarity, competition, choice, and flexibility.

The Canadian Congress of Labour, like the British unions it was modelled on, tried holding subject unions to their proper industry. For the same reason that good fences make good neighbours, that worked well in the heyday of industrial organizing. Miners in Cape Breton helped out steelworkers in Hamilton. Steelworkers helped a retail union

organize Eaton's. Everyone pitched in because no one had to worry about their own backyard.

Strict jurisdiction helped unions to become experts in an industry and co-ordinate national bargaining programs. It helped to hold leaders accountable for decisions affecting a common membership. That's almost impossible in a "conglomerate" union with members who have little in common. Most important, strict jurisdiction made unions fight employers, not each other.

But strict jurisdiction also kept out fresh blood, led to "hardening of the categories." That was most obvious with craft unions that operated in jurisdictions where factory workers didn't exist. The CIO was a rebellion against those outdated jurisdictions. Strict

jurisdiction decided by a central congress guaranteed a monopoly, and workers lost the right to pick the union of their choice. Isolated groups lost the chance to ride the coat-tails of a strong local union. In one-industry towns like Sudbury, for instance, the International Union of Mine, Mill and Smelter Workers was a haven for waitresses, hairdressers, and retail workers who could never have organized on their own.

The CCL also created outdated jurisdictions. It "grandfathered" chemical workers to a coal miners' union because chemicals were once made from coal. This jurisdiction was obsolete in the petrochemical age. It denied the interlocking ownership across the energy and chemical field and the principle

THREE-THIRDS BELOW THE SURFACE

of one union for one employer.

The CCL's was the worst system, except for the rest. U.S. labour emphasized strong individual unions that took what they could get and gave little power to central bodies in charge of a labour movement. Autoworker and CIO president Walter Reuther told the Oil Workers in 1951 that "our boys" eat pickles and "our girls" use makeup, so the United Automobile Workers organized pickle and makeup factories. Oil workers should organize any industry that uses oil, he said. The Teamsters had the same approach and organized any buildings that had wheels or walls. Every union for itself.

These tendencies formed the backdrop of the 40-year fight to find a jurisdiction for energy and chemical workers.

At a tender age, Oil Workers had to decide what they wanted to be when they grew up. The oil business, the very model of a high-tech, low-touch industry, offered no prospects for union growth. The entire refinery hiring list couldn't produce a decent-sized local of auto or steel workers. Yet, in awarding jurisdictional turf in 1949, the old boys' network in the Canadian Congress of Labour insisted the Oil Workers stay on their own side of the industrial tracks, well clear of their oil-industry cousins in the energy and chemical field.

Seven well-entrenched unions kept the Oil Workers boxed into their narrow pigeonhole. District 50 of the Mineworkers, favoured sons of the CCL because of the miners' role in launching industrial unionism, had the franchise for gas and chemical outfits once based on coal byproducts. That froze the United Gas, Coke and Chemical Workers out of the Congress — an exact reversal of what happened in the United States — but it didn't freeze Gas Coke out of the Sarnia area, where it had a foothold in the chemical companies created by the war. The International Chemical Workers had the backing of the Trades and Labour Congress, the original Canadian labour federation that was tied to the craft-based American Federation of Labor. In Quebec the Confederation of Catholic Workers accused international unions such as the Oil Workers of preaching materialism, atheism, and class conflict. Public-sector unions earmarked public utilities that marketed energy as their jurisdiction. If oil travelled by pipeline, labourers' unions sewed up the construction, often through "sweetheart agreements." The trans-Canada pipeline, for instance, was built by Hod Carriers, after the union's Chicago headquarters signed a no-strike agreement termed a "slave pact" by the Oil Workers. Finally, if oil travelled by truck, the Teamsters drove away with it.

The Oil Workers sized up their elders and decided they wanted to be an across-the-board energy and chemical union when they grew up. Since they couldn't beat all those unions, they joined them. That's how today's Energy and Chemical Workers Union was created: through a series of mergers — a union version of arranged marriages — and raids — a union version of elopement.

That expansion-team approach to recruitment made Neil Reimer the bad boy of the labour establishment for most of his career. It brought him into conflict with some of the most strong-willed and colourful labour leaders of his time. It put the puny Oil Workers in the thick of some of the most dramatic internal battles in Canadian labour history. It forced a debate on one of the thorniest issues of union structure, the issue of jurisdiction.

Reimer, for instance, had a European theory of jursidiction. He believed that unions should be generic, should represent entire realms of economic activity, such as transportation and communications or the whole energy and chemical complex. Pat Conroy, leader of the Canadian Congress of Labour, had a British theory of jurisdiction. He believed that unions should represent specific industries — steel or auto or chemical. Leaders of most strong unions had a U.S. theory of jurisdiction. They believed they should represent the workers they wanted to represent.

Growing up with mergers also made for a very different type of union. In the union-rush days of the 1940s, workers joined the first union to hand out cards, as long as it was with the magical CIO and promised to stand up against the boss. The Oil Workers and their successors came on the scene later, so they had to try harder and put stock in their "differentness." To meet the competition, they asked workers to choose them on the basis of bargaining strategies, service, or a record of membership control, Canadian autonomy, and community involvement. In this self-conscious process, the accent was on unionization, not organization. Emphasis shifted to leading, not reacting.

The demands of this milieu produced unions that were innovative, fast on their feet, ready and eager to seize the day and steal a few bases.

Early union leaders came from the Stand By Me generation. They fought for principles, but personal loyalty was the industrial-strength crazy glue that held them together through the constant barrage of organizing and bargaining drives. When the chips were down, fast friends had to count on one another.

McAuslane came from that generation. When he was with the Canadian Congress of Labour in the 1940s, he was first and foremost a loyal footsoldier of Congress secretary-treasurer Pat Conroy. He stuck by Conroy through thick and thin, and Conroy stuck by him through warts and all, their old buddy Harvey Ladd says.

From 1949 to 1951, the two friends were ripped apart by a seemingly petty issue that had to do with bog — with the accumulation of eons of rotting seaweed and extinct sea monsters. Whether this matter came to the surface as coal, gas, or oil was a quirk of millions of years of history. Whether the workers who tapped and refined the gas were Mine Workers, Oil Workers, or Gas Workers was a quirk of a few years in the late 1940s. The Oil Workers, with less than fifteen-hundred members and less than five years' standing in the country, gave the mighty Mine Workers and CCL gas pains until the Oil Workers won the right to represent those who handled the liquid bog. And when the glue that held McAuslane, Conroy, and their old comrades together came undone, the basic structure of the labour movement was irreversibly changed.

Conroy's labour ideals — like those of McAuslane — came from the depths. He had spent his teen years in the coalpits of Britain. He left for Canada, went underground in Alberta, and in 1919 emerged as a major leader of the coal miners' rebellion in Drumheller, Alberta. When that strike wave was defeated, he soapboxed for unionism across the United States, put himself in the midst of some old-time free-speech fights, and often got carted off to jail until the town sent him packing. Early in the 1920s Conroy came back to the Alberta coalfields to find a virtual state of civil war between hard-pressed miners and hard-nosed operators. He was, relatively speaking, a moderate, and stood by Mine Workers' international leaders who bore down on headstrong militants through the 1920s and 1930s.

In 1941 Conroy was everyone's choice for CCL secretary-treasurer. As in Britain, the Congress president was a figurehead and real power lay with the secretary-treasurer. The industrial unions counted on Conroy, a private-sector unionist with a reputation for supporting international unions and shop-floor militancy, to checkmate CCL president Aaron Mosher, a public-sector railway-union leader who distrusted both U.S. unions and strikes.

Once a CCL leader, Conroy tried to turn it into an independent Canadian union centre, not just a beach-head for the U.S.-based CIO. He wanted it to be a forceful presence in its own right, not just a creature of its affiliates. McAuslane was his hatchet man, detested by those who wanted to cut Conroy down to size but backed by Conroy to the hilt.

Conroy met his match in Steelworkers' leader and Canadian CIO leader, Charlie Millard. A deeply religious man, known as "Christian Charlie," Millard saw the socialist and labour movement as part of the same field-force that had driven the merchants from the temple in Jesus' day. With manic energy, inspired vision, and an outstanding ability to spot talent and give it free rein, Millard turned the United Steelworkers of America into the premier union of its day. As a top officer of the CIO, he also handpicked the leaders of several other unions. He was not a man to mess with. Though Millard had no qualms about cleaning out the brethren in all-night poker games, he was a stickler when it came to boozing and cursing — which just happened to be two of McAuslane's most prominent habits. According to Harvey Ladd, when Millard decided he wanted the drunkard fired, Conroy looked Millard in the eye and told him he'd rather have a tippler on staff than someone drunk on their own power and ego.

Though Conroy and Millard disliked each other, they held their feelings in check during the war and worked together on a number of projects. They lobbied behind the scenes to increase CCL independence from the U.S. CIO. Both insisted that the CCL be seated as an independent body at world gatherings. Both twisted arms to have CCL dues paid through Canadian rather than U.S. union headquarters. Both favoured a hefty dues increase to boost the CCL operations. For Conroy, these steps were the cornerstone of the Congress he wanted to build. For Millard they were the capstone. Conroy wanted a central and unitary government of labour in Canada, with the CCL at the reins. Millard wanted a federation, one expression of a community of labour communities, with freedom of action for each union. This was neither the first nor last debate of its kind in the country. Alas, it was a debate that never took place, because personal feuds got in the way.

To bolster his position, Conroy kept the independent locals that were chartered during the 1940s shut off in a sort of Congress bullpen. Originally, the Congress organized and serviced these locals on a makeshift basis, until an international union came along and claimed them. The Congress provided no specialized negotiators, no service reps, no strike fund. The locals were kept as isolated units, rarely grouped with an eye to co-ordination, strength, or permanence.

After the war, these independent locals became Conroy's political and financial lifeline. The lion's share of their dues went to the cash-starved Congress. They paid the salaries of Congress staff and provided a political machine for Congress leaders. Ironically, the weakest and most dependent locals became the mainstays of Conroy's strength and independence from the large international unions that also belonged to the CCL. Conroy's strategy, his long-time associate Howard Conquergood once told Reimer, was to enlist 30,000 Congress members in chartered locals. He needed that many to hold the balance of power.

According to Irving Abella's *Nationalism, Communism and Canadian Labour*, the internationals disliked this arrangement because they were "greedily eyeing" the thousands of dues-payers in CCL locals. The greediness was just as much in the eye of Abella's beholder, Pat Conroy. As the international unions in Canada saw it, Conroy was playing out of bounds. He forced them to pay dues to a Congress that turned around and competed with them in organizing. Moreover, Congress locals rarely went on strike and tended to set poor standards in their area.

After McAuslane left the CCL staff to head the Oil Workers in 1948, he agreed with these complaints. He was at odds with Conroy and his new organizing co-ordinator, Henry Rhodes. The Congress promised to transfer gas workers in its National Union of Gas Workers over to the Oil Workers. The industries ran parallel and wage rates should be parallel too, McAuslane thought. But he kept bumping into trapdoors. He came to speak to meetings, but no one showed up. The Gas Workers kept complaining about the Oil Workers' higher dues. Transfers kept getting postponed for another round of bargaining. McAuslane attacked Rhodes for breaking Congress policy. He spread gossip that Rhodes had a flame in Windsor, and used Gas Worker meetings there to keep it stoked. Privately, he saw the hand of his old boss, Pat Conroy, putting in the fix.

Conroy was exhausted by Congress infighting. He saw no end to it, foresaw no way to keep the aggressive CIO-affiliated internationals from throwing their weight around, calling their own shots on jursidiction, and turning the Congress into a figure-head. He decided to draw the line and dare the Steelworkers to cross it. That turned out to be Conroy's tragedy. Not that he got steamrollered by U.S.-based unions, as Abella argues, but that his timing and issues didn't clarify the conflict at hand.

In 1951 Conroy backed Sam Baron of the United Textile Workers of America for the CCL executive. Baron was an unsavory character but he was Conroy's choice. The Steelworkers ran a candidate against him: the strongest union in the Congress taking on a stand-in for the strongest man in the Congress. Baron lost, and delegates at Steel tables whooped for joy. Their badges and kits were still swirling in the air when Conroy stomped to the front, stated that he no longer represented the Congress, and walked out of the labour movement forever. McAuslane walked out with him, giving up his Congress vice-presidency. Baron later switched to the AFL's textile union, fired militants from his staff at the behest of Duplessis and his International, and started raiding CCL unions. The grudge match degenerated into unpredictable acts of personal loyalty and treachery.

None of this fazed the Gas Workers. Rhodes's favourite Sarnia local kept its 34 members in the Congress until 1954. It wasn't until 1968 that the last CCL local switched to the Oil Workers. To celebrate this end to an era, Reimer and his buddy, Buck Philp, tied one on, pledged to quaff one drink for every one of their 120 new members. "We made a good stab at it," Reimer says, though his memory of the evening is understandably not clear.

According to Abella's influential study, Conroy's resignation sealed the fate of independent Canadian unionism. The authority of U.S.-based internationals "triumphed over the nationalist aspirations of the CCL," Abella writes. But the issue was not posed that squarely. The large internationals, for reasons that had little to do with their U.S. connections, simply rejected a central body that meddled in bargaining and organizing.

The smaller unions that needed a strong labour central to enforce all rights equally, to lay out objective blueprints on matters such as jurisdiction, were the losers. For energy and chemical workers, it meant close to 30 years of coping on their own with the results of hasty, dated, and arbitrary decisions made in the early 1940s.

It's no accident that the Oil Workers first took on the outlines of an energy and chemical union in the West. The prairies have always been a place where Eastern distinctions fall flat.

Pioneers couldn't afford to stand on tradition. A small population in a large, tough, and lonely land couldn't waste time on distinctions that didn't work. The United Church, formed in the mid-1920s, was a product of the West, where ancient Protestant jurisdictions meant little in isolated communities seeking a more Christian world. The political jurisdictions of Liberals and Conservatives also had little meaning in the West, and protest parties regularly shut out both parties from the 1920s on. Westerners didn't get uptight about the jurisdiction between private and public ownership, and governments of all stripes tried out public telephones and auto insurance. In the world of labour, the One Big Union attacked the craft divisions born in the East, and won over Western workers in 1919, long before the CIO started in Ontario. The Oil Workers got their first breaks in the West too.

In 1952 Reimer stole Saskatchewan power workers away from the Canadian Congress of Labour. McAuslane covered for him, claiming that power workers were "oil workers on telephone polls." Reimer concocted a more elaborate justification, claiming that Sask Power would soon sell natural gas. That didn't wash with the Congress jurisdiction committee, headed by the Oil Workers' arch-rival, Silbey Barrett. The committee wanted to measure the impact of gas distribution on the workforce before it agreed to the transfer. Unless a majority of its workers actually handled gas, Barrett ruled, Sask Power wasn't a gas plant. McAuslane didn't agree with that paint-by-number approach. He told the Congress executive gruffly that "if other unions could do it and get away with it, he could not see any reason" why he couldn't too. The Congress hauled up the Oil Workers on charges of violating its basic principles.

Seen from Congress headquarters in Ottawa, the dispute centred on rules of jurisdiction. Seen from the powerhouse, highwires, and meter wagons of the workers of Sask Power, the problem was finding any union with the guts to take on the union-backed CCF government.

Power workers had been shopping for a union since 1949, when they bolted from the Saskatchewan Civil Service Association, claiming it was weak-kneed in bargaining. The Congress promised better service and a tougher stance, so the power workers signed on as an independent CCL local. The Congress didn't hold up its end of the bargain. In 1950 and again in 1951 power workers asked for authority to strike and got turned down. The CCL didn't want to embarrass the CCF government, labour's own. "The union committee sat in those sessions with their hands tied behind their backs" and had to accept whatever "peanuts" management offered, local leaders griped. The last straw came when the CCL ordered the government's striking bush pilots back on the job.

In 1952 Reimer bumped into an old Regina sidekick, Chester Tatlow, in the Regina airport. Tatlow was the burly power player of the local, so Reimer asked where he was going and quickly decided he was going there too. "We have to find a home someplace," Tatlow said, as the plane lifted off. "It can't go on like this with the CCL. They won't let us strike." He asked Reimer's opinion. Reimer suggested that Tatlow's local join a union with higher wage rates, a new union that could pull them up, not drag them down. "I'll never stand in the way of your fighting the boss," Reimer promised.

Tatlow set up meetings for the Oil Workers to make their pitch. Of 700 eligible voters, 659 voted to transfer. Of such majorities were new jurisdictions made, however Silbey Barrett's CCL committee counted the numbers. The move put the Oil Workers on the energy grid, a step past the oil and petrochemical field. It gave them the numbers to be something other than a skin and bones operation. It got them a new staff rep, a wiry scrapper named Cy Palmer, formerly with the CCL. It also put them in the spotlight of a province where labour political history was being made.

In 1953 Reimer organized three chemical plants located in Edmonton close to gas and oil supplies. He beat out District 50 of the Miners as well as Gas Coke and the International Chemical Workers. Their heads were stuck in the East. In the West what counted was not the size of a union or the piece of paper granting it title, but the fact, as noted by Bob Gallivan, manager of the Edmonton CIL at the time, that "The type of people we hired were the type of people we could induce from an oil refinery." In terms of work processes, petrochemical and oil refinery workers were kissing cousins.

Union leaders chewed out McAuslane for letting Reimer jump claim, but McAuslane dropped Reimer a line stating, "The question resolves itself as to who organizes the workers." In jurisdiction, as in other matters, possession was nine points of the law. Then the Congress executive laid down the law. Chemicals belonged to District 50, even if made from natural gas, it ruled.

Reimer, knowing the ruling would suffocate his union, wanted to take the Congress executive on at the 1953 convention. McAuslane, knowing the executive was counting on the votes of ten thousand Mine Workers in Cape Breton, warned Reimer to back off, forbidding him to even show his face at the convention. "We don't have a snowball's chance in hell of beating them," he said.

Reimer had it out with McAuslane. "We have no choice but to expand. We either go that route, or we throw in the sponge," he said. At any rate, Roy Jamha, then working as a Congress organizer in Alberta, had already geared up Western delegates for the challenge. The question resolved itself as to who had organized the workers. McAuslane agreed to make the appeal personally.

Silbey Barrett was classic salt of the earth: old salt, washed on him from the fishing boats he worked as a boy in New-foundland. He never went to school. Amidst a cast of rough-hewn streetfighters in the CIO, Barrett stole the show. He was forever chomping on an unlit cigar, stooping forward after decades of working the narrow tunnels in Cape Breton mines. He might have been illiterate but his deep Newfoundland accent, shrewd intelligence, photographic memory, and mastery of the tall tale made for an unbeatable combination.

Barrett became best known for what his old office mate, former Congress staffer and Newfoundland Senator Eugene Forsey, calls "Silbeyisms." He told one employer, "A verbal agreement ain't worth the paper it's written on." He told another, "We'll look at this agreement claw by claw." He likened many trick moves to icebergs, which floated "t'ree t'irds below the surface."

Barrett started mining coal in Cape Breton in 1902 and took part in the nine-month strike for union recognition that lasted through 1909 and, as he put it, "19-0-10." In 1919 he was elected local president and international board member of the Mine Workers. Throughout the 1920s, Cape Breton was in a state of virtual civil war and when the government repression came John L. Lewis denounced the local union leadership. Barrett stuck by Lewis, the man he always called "the boss," and in 1937 Lewis appointed Barrett Canadian head of the Steelworkers Organizing Committee. In 1945 he appointed him director of District 50, the Miners' catch-all subsidiary. As a member of the Congress executive from its founding in 1940, Barrett kept a sharp eye out for union competitors, and made sure the Oil Workers' jurisdiction was sealed tight as a drum in 1949.

McAuslane and Barrett, two peas in a pod, were long-time friends, travelling companions, and sparring partners. They gave jurisdiction the same respect they gave rules on grammar and outdated labour laws. McAuslane knew nothing about petrochemicals and cared less. In moments of great vision he insisted the whole world was made of chemicals. Barrett was a one-man One Big Union who included sailors and barbers in his District 50 jurisdiction. While travelling through Woodstock, Ontario, the two leaders stopped to survey a quarry represented by the Oil Workers. "You're supposed to be a great union man," Barrett snarled. "What in hell is an oil workers' union doing with them miners over there?" McAuslane shot back: "Silbey, when they dig deep enough, they might find oil." The joking became dead serious at the 1953 CCL convention.

McAuslane launched his appeal, stressing that he came to praise the executive, not to bury it. He said he appreciated that the executive erred only because the industry was so new and complex. Massacring the notes prepared by Reimer, McAuslane gave a lecture on the special nature of the chemicals that came from gas and oil. The Oil Workers came to Canada thinking they would have a broad jurisdiction, he said. The union was young and vigorous, and should go with the flow of oil and gas when they come out of the ground.

Barrett, for his part, praised the executive for its common sense. Then he raked the Oil Workers over the coals for having hired a chemical engineer to trick delegates into thinking that a chemical wasn't a chemical. Working the loyalty line, he reminded delegates that District 50 had organized chemical workers before the Oil Workers were even heard of, and that workers shouldn't be divided now.

The Congress president, Aaron Mosher, lent his authority to

57

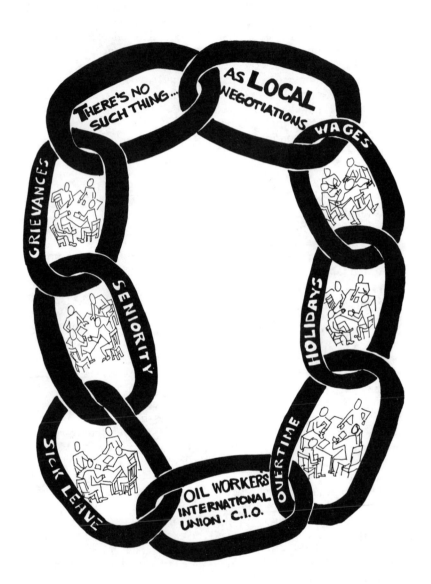

uphold the executive. Chemical plants remain chemical plants even if they use gas and oil, he said. If we followed Oil Worker logic, sheepherders would have jurisdiction over textiles, and loggers over construction, he declaimed.

Behind the scenes, where three-thirds of the votes were lined up, organization, not speeches, made the difference. The West was well represented. For two years, aware that his competitors felt safety in numbness, Reimer had directed Oil Worker activists to lead labour councils and federations so they'd have influence beyond their numbers. Roy Jamha, on temporary Congress staff, whipped up the mutiny against his bosses. "You were working on people as fast as you could get to them," he says. He stressed membership control in the Oil Workers and exposed its absence in District 50, which had no voice or vote at Mine Worker conventions. He broadcast Barrett's defence of the reason why he was appointed, not elected. "The Pope don't have elections," Barrett had claimed.

The Congress wasn't prepared for the groundswell. The Oil Workers won the right to organize the petrochemical industry. Jamha lost his Congress job but was hired by the Oil Workers. Barrett quit the Congress the next year, denouncing it for limiting his jurisdiction. From 1954 on, raids on District 50 chemical plants had Congress blessing and Barrett's union began to die the death of a thousand cuts. "The 1953 convention decided which union would live, and which would die," Reimer says.

BAD BLOOD, TIRED BLOOD

By the 1950s, the Gas,Coke and Chemical Workers had fought their way to the bottom.

Despite massive financing from the CIO, with its vested interest in bolstering a commando operation against John L. Lewis's District 50, and despite a crack team of organizers who enlisted thousands at a time in the booming war industries, the U.S. union couldn't manage its own growth, says former Gas Coke leader Tony Sabatine. Born of bad blood, bred for blood-letting, the union couldn't roll with the punches of everyday internal differences. In search of a new external enemy, Gas Coke became the first U.S. union to adopt anti-Communism as its war cry and witch hunting as its method of internal debate.

As the union's membership decreased after the wartime boom, the infighting became vicious. When Charlie Doyle, a British immigrant to the United States and one of the union's best organizers, went to a meeting in Windsor, Ontario, his union opponents tipped off border guards that he was a Communist alien. He was deported.

The Canadian region of Gas Coke got off to a good start in 1943 when it organized Polymer, the crown corporation making wartime synthetics in Sarnia. Then it scored at giant Cyanamid plants in Niagara Falls and Welland. But after the war it witnessed a major defection. The original staff rep and executive board member for Canada was Bill Edmiston, a Communist Party supporter. Before the Gas Coke red-baiters could get him, Edmiston jumped ship and went to the International Chemical Workers. Unhappily for Edmiston, his new bedfellows fired him as soon as he had burned his bridges — and then transferred his locals.

In 1946 Canadians in Gas Coke held on by the fingernails. The International couldn't afford to staff the Canadian district. To protect themselves from raiding, says local leader Russ

CHALKING UP THE MIRACLE INDUSTRIES

Gillespie, they temporarily gave themselves over to the United Automobile Workers for "protective custody."

By 1954 Gas Coke was still small, still stuck in the chemical belt of southwestern Ontario. It was cut out of the CCL, which honoured District 50's claims to represent the chemical workforce. Its top leader, Aubrey Bruyea, was set in his ways, had no delusions of grandeur, no desire to play on a bigger field. A founding member of the Polymer local, Bruyea knew only the most steady of relations with large and comfortable corporations. He liked his job and his title and saw no need or reason to change.

If the 1940s brought shake-ups, the 1950s brought breakups and shakedowns. The economic, political, and social system settled into steady growth. Labour leaders gave up on sudden breakthroughs and hunkered down for gradual bargaining gains and piecemeal organizing drives. They made new friends and broke off with old ones in sacrificial blood feuds.

The Canadian Labour Congress, formed in 1956, was one product of the decade's drift. There was no longer room for two competing union centres and two competing labour philosophies, so the craft and industrial unions decided to sort out their differences and merge. The New Democratic Party, founded in 1961, was a creature of the same trends, spawned by the breakdown of Depression-era radicalism and the new support for moderate reform and a mixed economy reflected in the CCF's 1956 manifesto. Last but not least came the formation of the Oil, Chemical and Atomic Workers in 1955. The Oil Workers International Union and the United Gas, Coke and Chemical

Workers, both products of the meteoric rise of unionism in the 1940s, shed their old names, skins, and differences to create a new union to meet the test of new times.

By the 1950s Gas Coke was the problem child of the CIO. It couldn't survive without a hefty allowance from its parent, so Walter Reuther, president of both the Autoworkers and the CIO, offered a dowry to any union that would take Gas Coke off his hands. Two unions can live as cheaply as one, he argued. Speaking to the Oil Workers' convention in 1954, he promised to donate a hundred thousand dollars to get a merged union off the ground. Such a proposition gets serious talks going in the United States — but not in Canada. The Canadian Gas Coke conference in 1954 agreed to co-operate with the Oil Workers in bargaining and organizing campaigns, but rejected the idea of merger. Canadian leader Aubrey Bruyea "fought merger all the way down the line," says Tom Dillon, who was president of the Polymer local in Sarnia at the time of the proposed merger. Bruyea worked equally hard at avoiding Reimer and the necessity of making a direct refusal to hold merger discussions. He was never in his office when Reimer called, never returned his calls. It seemed that he was either dragging his feet or digging in his heels. Reimer decided he would have to light a few fires to ensure movement in the right direction.

In October 1954, after a meeting of local leaders of the Oil Workers and Gas Coke, Reimer assigned Cy Palmer to a scouting expedition. Palmer got hold of Bruyea but the reception was chilly. Bruyea said staff layoffs would be necessary if the unions merged. He didn't have time to help Palmer organize Sarnia oil workers. Palmer could use the Gas Coke office as a mailing address, but there was no space to work there.

Palmer attended a Gas Coke local meeting at Polymer and helped members plan for a local labour council. The council couldn't be formalized until a merger was formed, because Gas Coke wasn't allowed in the Canadian Congress of Labour. While speaking to the local Palmer cleared up, as he put it, "some of the misunderstanding which has arisen due to Director Bruyea's alleged misinterpretation as to how the merger would affect" Gas Coke members. He wrote back to Reimer that the meeting gave him a "hearty vote of thanks." In Windsor Palmer met an area council of Gas Coke members and picked up widespread complaints about poor servicing. Several delegates told him they'd support merger at the upcoming February convention. But Palmer warned Reimer that merger opponents still had "considerable scope in figuring out new angles to influence people against the merger" before the convention.

Reimer kept up the momentum by inviting Gas Coke members to an educational conference just before their 1955 merger convention in Cleveland. The Oil Workers' local of 39 members hosted the event for the huge Gas Coke crowd and asked Howard Conquergood to lead the sessions. One of the great labour eduators of all times, Conquergood chalked up a victory for merger as only he could.

In a movement full of larger-than-life personalities, Howard Conquergood was larger than them all. In 1937 he was a lineman for the Toronto Argonauts, helping them win the Grey Cup. He was 250 pounds of muscle, an all-round athlete who starred in gymnastics, wrestling, and boxing. But he wasn't a muscle-bound jock. In the 1930s he was a leader of the CCF youth movement. During the Second World War he worked for the Toronto YMCA to set up a major recreational and cultural centre for workers. Conquergood wanted to offer alternative ways of living to workers, to speak to their spiritual as well as physical needs. He organized sports programs, communal vacations, and private counselling. That project caught the attention of the Steelworkers' Charlie Millard, who was always looking for Y-trained leaders who would offer labour alternatives to the saloon.

Conquergood was a standout on the picket line — a great place to throw weight around in a contact sport. He'd stick out his chest and stomach and dare the cops to shove him aside. He was arrested at least seven times for picketing activities. Indeed, legend has it that he once physically prevented a train from crossing a picket line.

Conquergood trained thousands of strikers to take on the tasks of running a successful strike and most of today's strike manuals are lifted from his courses. He taught workers to set up soup kitchens, canteens, women's auxiliaries, and "good and welfare committees" to help out members in dire straits. In difficult times, when workers saw the need to rally as a community, they valued Conquergood's passion for the union as a community resource and a training ground in town-hall democracy.

After 1951 Conquergood headed the CCL's department of education and welfare. There he developed the same approaches to adult education that would later be "pioneered" by literacy experts. "He taught people to think, not to follow, which didn't make him popular with all leaders," says Harvey Ladd, a long-time friend. "His whole idea was that an educated working class would be a militant working class." He opposed formal instruction. He promoted self-education, learning as empowerment, not training. He got people to solve problems in brainstorming sessions so they could learn to teach themselves. He convinced several unions to hold conventions where members could work through issues by sitting in discussions around a table, rather than simply raising their hands in a large hall after listening to a bevy of orators. For Conquergood, teaching unionism was the same as teaching citizenship.

63

OIL REFINERY IN SARNIA IN ABOUT 1950

He was also "a certified clown, character, exhibitionist and ham," Ladd says. During a demonstration in Ottawa Conquergood did a handstand on top of the Peace Tower. At one banquet he lit a two-foot pipe with a blowtorch. At another he walked down the aisles on his hands, with a top hat balanced on his feet. At yet another he let a 90-pound woman flatten him in wrestling. A notorious eater, he'd down 12 sundaes in a row just on a side bet. He had no problem devouring three steaks, which he always ate rare, and tended to chase them down with a whole pie. Conquergood took his joking seriously. What the union members didn't realize, perhaps, was that he used gags to trick them into staying for a Saturday-night session on parliamentary procedure. The tragedy of his reputation was that few people saw past his size or through his antics.

A big man, Conquergood took a special interest in small unions, which lacked full-fledged educational programs. He introduced many Oil Worker locals to the basics. Reimer frequently called on him to help members from small shops group into composite locals that could become self-reliant. Quite apart from that, Conquergood took a shine to Reimer, considering him one of the up-and-comers. He jumped at the chance to help out with an educational that might promote the Gas Coke merger.

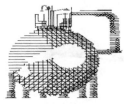

The educational was called to order at 9:00 a.m., February 5, 1955, at the Vendome Hotel in Windsor. The Oil Workers had five members in attendance, Gas Coke had a hundred and fifty. Conquergood explained the purpose of the session: to develop a better idea of the merger proposal that would soon be presented at a joint convention of both unions. He grabbed some chalk and asked everyone to think of reasons *against* the merger. All morning, he filled the blackboard. Anxieties and concerns came out into the open.

In the afternoon Conquergood covered the reasons *for* merger. "Let's go through this list. Why do you think that? What's the real problem here? How can we solve that?" He just asked the questions — the workers gave the answers. By the end of the day they were unanimous for merger. Bruyea and other Gas Coke leaders were at the back of the hall and when Reimer walked over to shake hands goodbye they turned heels, leaving his hand stranded in the air. That night Reimer and Conquergood took the train back to Toronto. Conquergood ordered a "blue, blue steak," and Reimer picked up the tab.

Later that month, delegates at a joint convention overwhelmingly agreed to merge. On the last night of the convention Reimer and Bruyea stayed up late discussing details of the transition. After a few hours' sleep, while it was still dark, Bruyea signed out of the hotel, left a brief resignation letter, and returned to private life. That day Reimer took a plane to an Ottawa meeting of the CCL where the Oil, Chemical and Atomic Workers was awarded the complete jurisdiction.

In Canada, the merger united 4,628 members from the Oil Workers with 2,500 members from Gas Coke. Reimer of the Oil Workers would be director. Seven years earlier, Oil Workers had been told to bury their talents in the oil fields. Now, under a new name and with a new organization, the union laid claim to what the *Union News*, its new paper, called "the twentieth century miracle industries."

The anti-climax came a year later, with the formation of the Canadian Labour Congress. To keep the peace the jurisdictions of all partners in the new labour central were frozen. Two groups were licensed for chemical and related industries: the International Chemical Workers, formerly from the craft-based Trades and Labour Congress; and the Oil, Chemical and Atomic Workers, formerly from the industrially-based Canadian Congress of Labour. It would take more than 20 years -- and more than a chalk-talk -- to bring them together.

TEXTBOOK STRIKES

There was a textbook quality to the strikes that took place in 1958 and 1959. It started in 1958 with a long hot summer of industrial unrest, underlined by the absence of any cold beer because brewery workers were on strike. Heavyweights like Inco and Stelco took on the biggest union locals in the country.

In 1959 the anti-union drive got ugly. At one end of the country, B.C. premier W.A.C. Bennett imposed tough new laws against union picketing. A leader of B.C.'s unionized fishers was jailed for writing, "Injunctions can't catch fish, cut logs, nor ... build bridges." At the other end of the country, Newfoundland premier Joey Smallwood outlawed the loggers' union and drove it off the island. In Montreal the CBC left French producers, among the first professionals to unionize, to cool their heels on the picket lines indefinitely.

Managers held to the Canadian Manufacturers' Association chant: "Hold that line in '59." More than that, they all followed a pattern. There was an uncanny resemblance between the tactics of Canadian employers and the Mohawk Valley Formula, a technique invented by Ohio employers in the mid-1930s when they still thought it was possible to drive back the CIO.

The Mohawk Valley Formula is a timeless blueprint for one-minute strikebreakers. Brought to public attention in Ben Stolberg's *Story of the CIO*, it sets out nine precise steps and tactics plus timetables for breaking a union. Steps one through five call for displays of police might against picket lines to make the union cause seem hopeless. Steps six through nine follow up with public announcements that the plant is once again running full-tilt with non-union workers.

Like any modern masterplan for total war, the formula focused on civilians as much as direct combatants, on psychological warfare as much as brute force. Based on the sound military maxim that an enemy within is more insidious than an enemy without, the formula featured a game-plan to turn people against one another, to break morale, and to discredit the union among its members and the community.

▼ PART I

▼ SECTION 2 / A LONG AND WINDING ROAD

▼ CHAPTER 3

POLY WANNA CRACK US

This formula was faultlessly orchestrated against the loggers of Newfoundland and nickel miners of Sudbury. In Sudbury the mobilization of strikers' wives against the union broke the spirit of the largest union local in the country. Elements of the formula were in evidence in every major strike in the country. Its tell-tale marks were campaigns to smear union leaders as outside agitators and to create centres of resistance within the union membership, especially within families. It poisoned the entire industrial relations scene. In the United States, employers never looked back because their unions never caught on. In Canada, chemical workers found the antidote.

With two years to go the 1950s lost their bloom, and union members lost their innocence. A drawn-out economic downturn gave unions their first close encounter with a concerted, nation-wide drive for concession-bargaining. The confrontation set off one of the most important, and least known, strike waves in post-war history.

For chemical workers in the newly formed Oil, Chemical and Atomic Workers, the period represented an especially rude awakening. They'd been raised on an industrial relations etiquette that closely matched a corporate paternalism. But once outside the comfort zone the young union taught its older siblings a few tricks about how to turn back a company offensive.

Looking like the original odd couple, Neil Reimer and Cy Palmer toured their new Southwest Ontario plants in 1955. A tall and husky farmboy, Reimer had a cool and calculated attitude. Palmer — short and scrawny, about 110 pounds soaking wet — had a personality

67

trained to overcompensate. Reimer assigned Palmer to service the old Gas Coke locals because Palmer was so ornery. "Since you enjoy fights so much, I'm sending you where they need some," he told Palmer. Reimer also wanted staff and members of the former two unions to mingle, to give up their has-been identity in a true merger.

No one looked down at Palmer for long. Bill Maugham was president of Gas Coke's Dow local in Sarnia and he says that when he first met Palmer, "He looked about seven-foot two-inches to me, he had such control and mastery of a meeting." To Maugham, any union that had a guy like Palmer seemed like a good union to join.

Employers found Palmer to be obstinate and cocky but meticulous about detail. He had worked for the Saskatchewan telephone system after migrating from England in the 1920s and had soon become the company's top splicer, weaving tiny cables through intricate switchboards.

In the 1940s Palmer started working for the CCL, where he proved to have an itchy trigger finger as a negotiator. He couldn't complete a session without at least one tantrum that went something like: "Had you calculated to offend us and inflame the situation, you could not have done a better job." He found it hard to accept any settlement, and usually called in Reimer when a deal got down to the fine strokes.

By Sarnia standards, Reimer and Palmer were a tag team from another industrial relations planet. They practised the old hard cop, soft cop strategy. Tom Dillon of the Polymer local marvels at the one-two combination. "Cy was an old rough and ready guy. The companies got so they didn't like him very well, because he was too tough on them. Then Reimer'd come in, and talk nice to the company with firm conviction in his voice, and the first thing you knew we had a settlement. It worked wonderful."

Ontario companies didn't know how to take their measure. They were used to the theatre of industrial conflict, not the real thing. According to Dillon, in the past the companies had figured Gas Coke wasn't "strong enough to have a strike, and didn't have a leadership that wanted a strike. So they just gave us what they felt like." Strong words and bluff were a ritual, not a warning.

According to Tom Towler, who was president of the Polymer local and later a Polymer corporate vice-president, the employers gave the sign "to take a strike vote, there's another half-cent there.... But there was nothing in terms of the other kinds of things that affect a workplace. It was just a very sterile type of relationship." Likewise, at John Wyeth in Windsor, Gas Coke's *United Chemical Worker* reported on May 26, 1952, that one-day negotiations had become "a habit ... indicating the fine relationship between the company and the local union."

With Reimer's arrival those days were over. "I had to teach the workers and managers that a strike vote's for keeps, not just a scare tactic." Managers smiled politely and held onto their wallets. They'd heard this before.

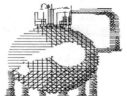

A series of workplace tours hardened Reimer's mind.

In Windsor he and Palmer soft-shoed through the Wyeth factory where "They played soft music at the beginning of the day," Reimer says, "and as the workers got more tired, they turned on the martial music." On one line Reimer overheard a foreman dressing down a woman who was behind in her count. "Now Mabel, did you just get your period?" Reimer turned to Palmer. "Strike 'em," he said.

At nearby Sterling Drugs, the company president kept them waiting in the boardroom. "I guess if he can't be on time, you'll have to strike him too," Reimer said on the way out.

A local strike wave followed the tour. "We had to strike to show we were militant on the workers' side and to gain recognition as a serious force from the industry," Reimer says. "They had a terrible relationship, and the fastest way I thought we could straighten things out was by having a few strikes."

For Reimer, a few nice words and a strike worked better than a few nice words to establish a respectful relationship. The relationship itself, not the conflict in bargaining, was the issue. Reimer didn't think that strike threats worked to resolve simple bargaining differences. "Maximum bargaining strength comes just prior to a strike, when the company's deciding how much to pay to avert a conflict. Too few union people understand that's why companies are prepared to make concessions. If companies think they face a strike anyway, they hold something back to end it with. So, as soon as the strike starts, the union's bargaining power goes down sharply, because the company decided to take you on. Then there's only one issue, the survival of the union. The issue changes the minute a strike starts."

The issue changed quickly in the Sarnia area.

In September 1955, Fiberglass workers voted 300 to 6 to strike for more money and recognition of the steward body. Two months earlier the five-year-old local didn't have any stewards. Stewards are the marshals of labour relations, officers who keep labour relations from turning into a wild-west conflict between contracts. A strike to recognize stewards was a strike for union law and order in the plant.

Reimer organized a Conquergood-style strike, designed to build a community, not just wait out the company. Unionville, an instant suburb built opposite the plant, had a canteen, first-aid station, barbershop, and its own paper, *The Strike News*. When a local alderman threatened to shut down Unionville, strikers rallied downtown shopkeepers with a petition campaign that stressed the importance of higher wages to the local economy.

Unionville had a town hall where strikers' wives met, "to the end that we may take a more effective and active support of our husbands and thereby in our own interests," they said. Merry Reimer had showed her husband the importance of women's auxiliaries in the 1940s. She thought strikes were lost in the bedroom, not on the picket line. She also wanted a way to plug homemakers into the larger social and economic world. By 1955 standards, this form of all-in-the-family unionism was considered ultra-feminist.

After 74 days the union won a "smashing victory" that set new standards for the industry and the area. The strike paid dividends in 1957, when the newly respected local bargained the highest rates in the oil and chemical industry. Local unionists spread the news by passing out leaflets at Imperial Oil and inviting workers to catch up by joining the union.

Bargaining at Polymer (later Polysar, then Nova) took place in a goldfish bowl. The town of Sarnia was highly possessive towards its crown corporation, a crash project built to produce synthetic rubber for the war effort. It took 15 months from ground breaking to the first run of rubber. "From top to bottom, I know of no project any place where people gave as unselfishly as they did here," says Ralph Rowzee, an engineer who had become the company's vice-president and general manager by the late 1950s. Sarnia regularly voted Liberal, a political offering to appease the federal gods who had contributed so mightily to its existence.

In February 1957 the news spread that bargaining was going nowhere. Local papers carried scare stories about how the company had government backing for a hard line. Polymer blanketed the media with anti-unions ads. On Sunday February 10 every church in town began with a prayer against a strike. The union hall was open all day as fourteen hundred workers, a 91 per cent turn-out, cast their secret ballots. Reimer wanted to make sure that no one could accuse the union of "rabble-rousing" or pressuring workers at a mass strike meeting. The vote showed 92.5 per cent in favour of a strike if the company didn't settle in three weeks.

The company immediately sued for peace. The Liberal government appointed a high-level mediator, Eric Taylor, who reported directly to C.D. Howe, kingpin of the federal cabinet and czar of the economy. Palmer told the media that the appointment of a mediator was belated, as were the "pious statements" of the company.

Taylor got down to work. "He was a pretty strong personality," Ralph Rowzee says. "He represented experience in labour relations that was totally lacking in management at that time." Taylor weighed in on the union side. He proposed a staggering 142 changes to the collective agreement and a wage hike of 20 per cent. C.D. Howe reportedly told local managers who protested the rich agreement, "I don't pay you to disagree."

On March 5, the *Financial Post* sounded the business alarm. "Canada's chemical industry this week readied for a renewed higher wage onslaught from unions. Oil, Chemical and Atomic Workers have succeeded in leapfrogging wages at Polymer Corporation in Sarnia over similar rates in other major companies. The effects of this contract will spread widely and they'll be felt not only in chemicals, but also in another big Canadian industry — oil refining."

The victory was beyond the union's wildest dreams. Reimer started fretting about the next round. He knew this was too good to last, that the company would now have to take the union on to win back its credibility.

By 1957, according to Peter C. Newman's *Renegade in Power*, "Canada's boom had finally lost its momentum." Megaprojects had dried up. The baby boom had let up. High-tech plants weren't hiring — they already produced more than they could sell. Investment in new production dropped 25 per cent over the next five years.

The economy was bloated. Government and business tried to cure their gout with a purge. In 1958, Bank of Canada Governor James Coyne railed against Canadians living beyond their means and jacked interest rates to the highest level in 40 years. Unemployment shot up to 7.1 per cent, the highest in the industrial world. The Progressive Conservatives were now in power and Prime Minister Diefenbaker set a 5 per cent ceiling on all new wage settlements.

The OCAW was having none of it. The union proclaimed that it wouldn't accept any contract without a substantial increase.

The fighting back started at CIL's Edmonton plant in the fall of 1958. The company's refusal to even "change a comma" from the previous agreement provoked a strike. CIL followed all the cues of a developing nation-wide anti-union drive. It tried to keep production going by using tradesmen from construction unions. It encouraged workers to break the picket lines. It took pictures of all people active on the picket line. It appealed to the wives of strikers to get their husbands back on the job before it was too late.

The strategy backfired. Construction tradesmen defied their own unions and honoured the picket line. Only seven workers crossed the line. The rest collected strike pay from a special levy on all Canada's OCAW membership. Wives set up an auxiliary and gave lasting support.

Union staff rep Walter McCallum did weekly broadcasts on radio station CHED. He accused the company of being run by irresponsible outsiders. "One of the major problems," he said in one broadcast, "is the fact that all decisions are dictated to local management from the Head

71

Office in the East, by people who do not live here and who know nothing about local conditions."

In the end the CIL workers got an agreement that provided hefty wage hikes and protection from layoffs.

Sarnia was where chemical industry rates and standards were set. In 1958 Polymer started cracking down on unionists who got out of line. Lab workers had joined the union in the flush of enthusiasm following the 1957 agreement, and they went on a wildcat strike to protest the company's disciplining of one of their co-workers. The company threw the book at them and sued the union for $50,000 in production losses. Arbitrator Borah Laskin ruled in the company's favour in a precedent-setting case that allowed companies to use the grievance procedure to go after the union. Until then, grievances had been an outlet for worker, not company, complaints.

In 1959 the company demanded major changes to the 1957 contract. It wanted more freedom to contract out work to part-time employees, a company demand usually associated with the tough bargaining trends of the 1980s. The company also demanded fewer restrictions on job security and promotions — an especially sore spot for unions. Over and above any other merits, seniority promotes internal group solidarity and eliminates the benefits of "apple-shining," or currying favour with the boss, as the way to get ahead. That's why company demands around seniority are treated as veiled attacks on the union itself.

Polymer's Rowzee says, "We gave too much in 1957, and in the two years that followed, we figured we just had to take a position. I'm sure the union didn't believe that a crown corporation would be allowed to strike. We had to show them."

The union charged that the company was out to "gut the union." It was in no mood to give anything up. Negotiations settled nothing. The federal government sent in Eric Taylor again, but the Liberal mediator couldn't work his magic under the Tory government. Reimer refused to bargain with a gun at his head and said he wouldn't meet until the company had disbanded the "freedom clubs" that had appeared in the community and were trying to organize a back-to-work movement in the event of a strike. He called for an RCMP investigation of the clubs. The company denied any involvement in the clubs.

The strike began March 19, 1959, the first in the company's history, with eighteen-hundred workers on the outside. They built "Strikeville" to house the welfare and employment committees needed for a long siege. In contrast to a small town's traditional Chamber of Commerce welcoming sign, the banner at Strikeville announced: "Poly wanna crack us — Poly can't-a crack us."

Everything was clean cut. Union barbers offered free haircuts. "Helping keep families looking neat and respectable is all part of morale-building," the *Union News* reported. "We created a totally self-sufficient town within a town," Reimer says. Dave Pretty, a millwright in the plant, set up a garage to provide free car repairs for strikers. Strikeville not only had a marching band but also a navy, to swamp efforts by the company to land supplies by ship.

The freedom clubs purchased ads calling for a back-to-work movement. Families of strikers were confused. "The women in some households are the dominant factor," says local leader Russ Gillespie. "They didn't like it. They wondered what it was for. They didn't understand the issues." The union called a meeting to organize strikers' wives and six hundred women came out to the founding meeting.

The women didn't have time to waste on picket lines outside the plant. Direct action was more their style. Some of the strikers' wives worked in the office at Polymer, which was unaffected by the strike. Some of the strikers used their extra free time to go smelt-fishing. Rotting smelt soon found their way into the plant's ventilating system. That made it harder for supervisors to turn their noses up at the unlucky strikers outside.

The auxiliary denounced a woman who was reading freedom club ads on the radio, nicknaming her "Tokyo Rose." Women led a strike cavalcade through town. "To protect children's and our future, our men must protect their jobs," their placards said. Other placards announced, "We'll walk our legs to the knees fighting Poly policies" and "CMA says `Hold that line' So here we are."

The strike lasted 98 days, kept most of the gains from 1957, and added a nine cents and hour increase. The company got some increased flexibility in contracting out. It took a long time before the plant returned to normal. Local union president Tom Dillon didn't speak to plant manager Rowzee for years. The 60 or so members of the "freedom club" got the silent treatment for the rest of their working days. They had "a very lonely existence," Dave Pretty says. Years later, Rowzee and Dillon re-established mutual respect. "The company paid a good deal more attention to labour matters after that strike," Rowzee says. "There's been no strike since, so I guess both sides took the lessons to heart."

This was one of the strikes that brought the offensive by employers to a halt. It showed that the Mohawk formula unlocked genies that could be turned against their masters. Despite a few big wins, employers got an object lesson that unions were here to stay, including a young one like the OCAW.

STRIKEVILLE

POLY
Wana
CRACK'US

LOCAL·16·14
O.C.A.W.

POW

CR

ON
STRIKE

YOU
CANT
JO

LEAPFROG COUNTRY

The Canadian labour movement, like the country it comes from, has more geography than history.

The Canadian Labour Congress passes scores of motions for labour solidarity and kicks off dozens of nation-wide campaigns. It's probably the most resolutionary labour central in the world. But when delegates leave conventions, they're on their own. Canadian unions have no central policies, such as the solidarity wage policy in Sweden that binds all unions to fight for equal wage rates. The national centre is a figure-head organization. It doesn't have the resources or power to direct a national movement. Activists complain that the CLC stands on ceremony, but in fact it has nothing else to stand on.

Few national union directors are any better off. Over half the unions in Canada — most notably the Teamsters, United Food and Commercial Workers, and all the construction unions — have what's called a "business agent" structure. Power lies with locally-elected business agents who run their own show and keep the national director on the sidelines. Public-sector unions and the Steelworkers let their national leaders handle diplomacy, media relations, and an occasional make-work project, but locals, regionals, and occupational groups call the shots on key issues. Autoworkers and loggers have strong national leaders but do not have national memberships, a reflection of the fact that there are very few national industries in Canada.

Canada is set up so that workers and unions think provincially. Provinces and municipalities handle the meat and potatoes of politics, such matters as education, social policy, and labour legislation. Only a select minority of workers come under federal labour laws. Provincial labour boards set the rules for most workers, even those who work in nation-wide companies and industries. The boards are biased towards local bargaining. They insist that bargaining units have a "community of interest," defined in a narrow way. Unions have to organize companies one-plant at a time and even have to organize separate units for white-collar and blue-collar workers, full-time and part-time workers. Their "community of interest" in a national setting is denied.

This system leaves the economic elite with a monopoly on the big picture, and leaves unions with scrambled pieces of a jigsaw. Unions adjust through

GUESS WHO'S COMING TO BARGAIN

plodding bargaining methods that put a premium on "leapfrogging" (sometimes called whipsawing) rather than nation-wide mobilizations and breakthroughs. The first union gets a dental plan; the second union leapfrogs over its back to get a dental plan plus a raise; the first union waits its turn to leapfrog for a bigger raise, and so on.

The system gives employers the jump on labour. It allows them to fend off union blows one at a time. It limits the bargaining agenda to what a union can get on its own. It trains workers' sights up another union's backside. It fosters the pettiest form of solidarity, based on catch-up, not forward-thinking. All very good reasons for employer opposition to national bargaining.

Canadian history happens once in a while, in times of national emergency, sometimes during federal elections. The rest of the time, history is a more local affair. Few issues and fewer institutions bridge the country's many echo chambers: the solitudes of French-English, Maritime-Upper Canadian, West-East, North-South, rich-poor, white and non-white. The labour movement doesn't cross these barriers very often. As of August 1989, fewer than 5 per cent of union contracts linked locals across the country in the most basic of union activities.

The national bargaining program of the Energy and Chemical Workers, launched by OCAW in the 1960s, is an exception. Neil Reimer likes to praise this bargaining program as a contribution to national unity that few premiers' conferences can match. "A togetherness develops. The workers recognize that their welfare is linked to everyone else's," Reimer told the *Edmonton Journal* on August 15, 1975. "If workers understand this, I think there's less chance of balkanization in Canada."

77

Reimer likes to credit his immigrant background for that outlook. As an immigrant he was attached to the country he came to, not the region he was born in, he says.

There's more to his immigrant background than that. His dad had a passion for the national pastime of his homeland, the Soviet Union. During the Depression, when duststorms howled across their farm, father and son played barnyard chess 'till the cows came home. Reimer grew up with chess: with strategy, with playing the whole board, with deliberate moves that passed up smaller pawns to checkmate the king. The game proved to be an excellent preparation for the strategies of national bargaining. National bargaining meant that the energy and chemical workers got to make Canadian history, in the strictest sense of the word. Their strike antics in the 1960s made national headlines. Their strike victories set national precedents for wage parity, job security, and health and safety. Later, in the 1970s and 1980s, they used gentler persuasion to build on these gains in the face of runaway inflation and stop-and-go recession.

Some people say it's lonely at the top. That's because they don't know what it's like at the bottom. Lonely is Neil Reimer in the 1950s, trying to crash the party of world giants in the energy and chemical industry. As head of a union with a few thousand members strung across Canada, he wasn't invited. Long before the job-hunting guide *What Color is My Parachute* buried the myth that people work up from the bottom — people at the bottom get stepped on, not noticed — Reimer had figured out that it was easier to start at the top.

In 1954 he wrote the senior vice-presidents of Imperial and B.A. and offered to meet them in Toronto. He'd never met anyone higher than a plant manager before. The corporate heavies had never met anyone lower. Meeting a union director for a loose chat piqued the cloak and dagger side of their corporate imaginations. When "Soapy" Church — so named because he was so smooth — got his letter at the B.A. head office, he showed it to his assistant Ted Gaunt. "Let's go. But don't show that letter to anybody," he whispered. "Tear it up, and don't let anybody know that we got a letter from him." Imperial's vice-president told Reimer he'd prefer a phone call next time.

At the meetings Reimer brought up nothing particular. The meeting was the message, it put management on notice that Reimer's strategy was to go to the top. It was the only way he knew to get a handle on the paradox of bargaining in the oil industry.

The paradox was stated neatly by Jack Knight, international president of the Oil Workers union. Oil was a good place to start with new bargaining demands. "The panning is best where the gold's most plentiful," Knight told *Canadian Labour* in 1957.

But the wealthy and cunning oil bosses "are never guilty of being penny-wise and pound-foolish; they spend generously to build the delusion among employees that they do not need a union." As a result, Knight said, only one-third of the North American industry was unionized. "Thus bargaining always proceeds from a minority position, requiring the most careful planning and strategic application of such strength as is available."

Reimer sensed that national bargaining was the way through this paradox. National bargaining did three things at once, he told the Oil Workers' Canadian Council in 1956. It unified oil workers across the country, it anchored the drive to organize the unorganized, and it promoted outreach to other unions.

In this strategic plan, the last two points addressed the need for a centre of gravity in the union's policies and reputation. Basic steel set a high standard for the Steelworkers. The Big Three made the Autoworkers' reputation. Unions without a centre of gravity carried their membership as dead weight. They started from scratch each time they bargained. They had no targets, no sense of their due. That was how the International Chemical Workers and District 50 bargained. Reimer figured he'd get a leg up on them if he set refinery wages as the benchmark for a union their members could join. He also reckoned that a national bargaining drive could either expose, or spur co-operation with, company unions in the oil industry.

The first point was part of a strategy to cut through the charade of company paternalism in oil bargaining. Throughout the forties, fifties, and sixties, companies insisted that bargaining take place locally. The country divided itself logically into five regions, each with distinct challenges. Within a region, each plant had its own problems and prospects, the companies said.

Top management knew this was a crock. Bill Baker, then a senior bargainer for B.A. and later vice-president at Molson's, freely admits: "The pattern was set in Clarkson in those days and then the rest of the managers had to play the same tune. So only one [plant] was actively involved." Norm Devos, another industrial relations manager with B.A., says local managers could only strike a deal "if we had agreed identically to everything." If head office "agreed to everything, i's dotted and t's crossed, sure they could make a deal." Senior management's problem was that local managers "almost convinced themselves that they had the authority to negotiate contracts," Devos says. "That was always sort of an uneasy situation inasmuch as the plant managers always felt that they really had the authority to make a deal, but they didn't."

Local bargaining was a convenient fiction for the company. It skewered the union with a catch-22. The union couldn't hold out at one plant unless it organized the whole chain; otherwise, production would just be transferred to another location. But the union couldn't make the case to organize the whole chain unless it bargained nationally. That's because the industry rigged its offers to give non-union plants the same deal as union plants — only earlier.

Bob Kirk found that out in the 1940s, bargaining with B.A. at its showplace refinery in Clarkson, Ontario. "Everything we ever won, they gave to everybody else coast to coast, before we got it," the long-time president of the Clarkson local says. "We could tell we'd won a battle because word got back to us that others were getting it, so we knew next meeting we'd get it too."

The industry strategy had local unions chasing themselves in a circle, racing to keep up with what seemed to be management leadership in wages. Reimer's game-plan gave the industry more than it bargained for. It started as a prank on Imperial. Imperial used to call its "donkey councils" for a 10 a.m. meeting and announce the new wage rates. Reimer got a call from a Halifax worker, then raced against the time zone. By the time the 10 a.m. meetings were called in other parts of the country, the union had already distributed leaflets to each plant, announcing the news. "It sort of took the glow off their paternalism," Reimer says. "You were just a nasty bunch of bastards, that's all," says Devos.

The industry's paradox was that it was too smart by half. "They played into our hands, because they conditioned workers to think in terms of national bargaining," Reimer says. "By trying to resist the union, they built an environment that what one gets, the other gets. So it was easy for us to say: `Okay, we've got national bargaining; what we want is uniform rates.' I don't think they ever thought of the consequences of that."

At least one union activist had an inkling of the need for national bargaining as early as 1950. When Alex McKenzie, a stalwart of the New Westminister Oil Workers local, retired in 1950, he spoke to the *International Oil Worker* of his "dream" of "a master contract for all oil workers in the not too distant future."

The big hurdle was convincing union members to give national office the authority to bargain. That meant a major transfer of power within the union — from the locals to the centre. That meant each local would have to give up its bargaining autonomy. That took a lot of confidence in other locals, and in the national leadership.

Reimer knew enough to tread carefully. Three locals from Saskatchewan, where he had most influence, submitted resolutions for common wage hikes to OCAW's first Canadian conference in 1956. The resolutions, demanding raises from employers, not changes within the union, seemed innocent enough. They passed unanimously. Two other meetings that year opposed wage inequities in the oil and chemical industry and adopted common bargaining goals for the oil industry.

Reimer turned the resolutions into a field day for organizers. In November 1956, a hundred activists blitzed refineries in nine cities with leaflets asking "all employee groups" to join behind the bargaining drive. Organizing, or at least collaborating with company-based works councils, was built right into bargaining. In Sarnia, where non-union Imperial Oil held sway, the blitz made front-page headlines. "Every time we leafleted, up would go the wages," Reimer says. "Just the thought of us putting on a campaign among their employees was worth money."

Organizing was something unionists understood. Reimer put it front and centre in national bargaining. "If you kill the climate of growth, you seal the doom of the union," he told Canadian Council delegates in 1958. In 1960 the union organized K.C. Irving's refinery in New Brunswick, its first break into the Maritimes and first opportunity to put meaning into national bargaining. In 1963, when OCAW represented a little more than a third of the refineries across the country, Reimer invited leaders of 21 works councils to bargain jointly with the union. Managers sat up and took notice. As the October 1964 issue of *Executive* noted, the union scoreboard "reveals some cause for management alarm."

Bargaining habits were harder to change. In 1958 Reimer blasted oil locals. If you don't stick together better than in 1956, there's "not the faintest hope of success," his memo said. Reimer scolded, preached, organized. That was the limit of his formal authority. In 1963 — four years after a similar decision in the United States — the Canadian Council gave the director constitutional rights to direct national bargaining — on paper. The funds and structures were still missing.

National bargaining meant surrendering local autonomy. It required a level of trust that Reimer hadn't won. John Kane, then a leader of the B.A. local at Clarkson, suspected Reimer was "too cosy" with top management. "I felt he was wheeling and dealing behind our backs," he says. Later, when Kane joined union staff, a management negotiator asked for an off-the-record sense of union objectives. Kane turned to his clergyman for guidance. The preacher said it was alright, as long as workers benefited. "I understand now that there are approaches behind the scene that are essential," says Kane. "But when I was a rank and filer, I didn't see it."

Ron Duncan, who replaced Reimer as director from 1963 to 1967, couldn't enforce the discipline for national bargaining. Locals "enjoyed the freedom to go six ways from Sunday," he says.

The opposition wasn't anything he could grab hold of. Minutes of a 1964 oil bargaining conference reported that everyone agreed on "a feeling of need for a mandatory policy" but they were still "skeptical of a move in this direction." Duncan pleaded for honesty "as he felt that some of the delegates were hedging and that they had to put trust in each other and make their remarks openly so that we could formulate a policy that all were prepared to try and get passed in their local unions." Three weeks later Duncan wrote to Ray Davidson, the International's public relations officer: "Between you and me, I doubt very much whether the mandatory program will carry."

Events took over. In 1963 Shell bought out the last independent Canadian oil company. Foreign monopolies controlled all the industry. "The elimination of the independent companies has removed from the Canadian scene the lever that we have had over the years," Reimer wrote Knight. "The size of these corporations dictates a certain amount of economic weakness for us, even if we did have all the bargaining units organized." A brief drop in Canadian profits, less than 10 per cent of any company's assets, wouldn't budge head office. "The winning of a strike here would have to be predicated on inconveniencing the public," Reimer concluded.

In 1963 Reimer's logic formed just a premonition of the dramatic changes in union tactics that would be seen when national bargaining rose to the top of the union's agenda. As happens so often with unions, great opportunities are not so much grasped as thrust upon them. That's what happened in 1963, 1965, and 1969.

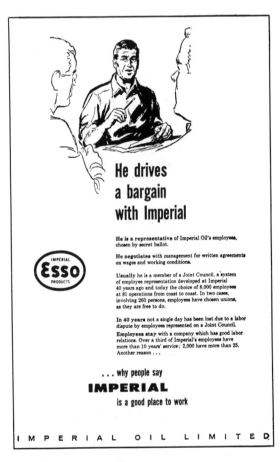

He is a representative of Imperial Oil's employees, chosen by secret ballot.

He negotiates with management for written agreements on wages and working conditions.

Usually he is a member of a Joint Council, a system of employee representation developed at Imperial 40 years ago and today the choice of 8,000 employees at 81 operations from coast to coast. In two cases, involving 260 persons, employees have chosen unions, as they are free to do.

In 40 years not a single day has been lost due to a labor dispute by employees represented on a Joint Council.

Employees stay with a company which has good labor relations. Over a third of Imperial's employees have more than 10 years' service; 2,000 have more than 25. Another reason . . .

. . . why people say

IMPERIAL

is a good place to work

IMPERIAL OIL LIMITED

COMPANY PROPAGANDA
VANCOUVER SUN, 1957

SMALL IS BEAUTIFUL

Local bargaining is favoured by Canada's system of labour relations. It's also favoured by the historical fiction of labour and radical movements across North America.

People came to the New World to be free, to be independent, to get authorities off their backs. The United States had its pilgrims in their Mayflower. Canada had its *coureurs-de-bois* in their canoes, paddling the country's vastness to put distance between themselves and France, which ran the fur trade as a centralized monopoly. "Jack's as good as his master" was the creed of the pioneers of the mid-19th century. The country's first labour organizers were travelling tradesmen, Johnny Appleseeds who sowed seeds of independence in the workplace. They didn't bargain collectively; they simply set a rate that good unionists enforced on their own. If the twentieth century belonged to Canada, it was because of its "last west best," the last place in the world where poor people could start up "on their own." That's how freedom was imagined.

In the 1930s, the radical Co-operative Commonwealth Federation was exactly what it said — a federation, not a centralized political party. The idea of a centralized political party controlled by Central Canada was exactly what militant Western farmers were out to destroy. It was in the "backrooms" of Ottawa, far from the "grass roots," where the forces of evil conspired. Elected CCF MPs were expected to sign a statement of resignation, which the local riding could act on as it saw fit.

Industrial unions of the 1940s and 1950s saw themselves as part of the same grassroots force. To take on the giants of industry, they organized all workers in a plant into one union. But they didn't bargain with their industry on a central level. Militants wanted to keep decisions close to the "rank and file," away from the "top brass" who might "sell them out." Local autonomy was seen as the hallmark of democratic unionism.

Slogans of the 1960s — "masters in our own house" or "for an independent and socialist Canada" or even "tune in, turn on, drop out" — all expressed a bias towards isolation, not interdependence, as the basis for liberation and self-expression. Today's activists prefer networking to party-building.

The equation of freedom with "freedom from" outside influence "immunizes" Americans "against all ideologies that require sacrifice of concrete personal interest in behalf of remote or abstract collective goals," Richard Flacks writes in *Making History*. The same could be said of Canada, where "small is beautiful" has long been the practice of popular movements.

83

ALL IN THE FAMILY

Owning everything in sight is what Irving is all about. "Not since the industrial imperialists of the last century," a *Maclean's* article stated on May 2, 1964, "has one man won and held dominion over people and wealth on the lordly scale of K.C. Irving." The magazine valued his personal empire of fuel, land, woodlots, hardware, pulp mills, public transit, and the media — he owned major radio and TV outlets as well as five of New Brunswick's six dailies — at $400 million. One-tenth of the provincial workforce was on his payroll.

Well over two decades later the Irvings still make their dollars go a long way. In February 1989 *Maclean's* rated the family fortune at $10 billion, making the Irvings the third richest non-monarchs in the world. They are the eleventh most wealthy family in the world. The largest conglomerate in North America, the Irving network of 300 companies owns a fleet of ships larger than Canada's navy, 3.4 million acres of prime timber, a shipyard, 14 planes, most of the province's media, the country's largest refinery, and a $2.5 million tax haven in Bermuda.

"We just run a small, family-owned Maritime business," Irving heir James Irving told *Maclean's*. The family business has stretched its dollars to prevent competition and to finance expansion through internal transfers of profits. According to Irving biographers Russ Hunt and Robert Campbell, "Each step in the integration not only eliminates a possible source of opposition or friction," it provides "the necessary capital to rejuvenate the company and assume control."

The refinery laundered both control and cash flow, according to Senator Charles McElman's testimony to a 1971 commission into media monopolies. Irving bought offshore crude at top dollar to justify high prices at the fuel-pump and lower wage rates in the refinery, McElman said. A nice way to make ends meet.

The air-tight system throttles competing sources of power, whether they are other businesses, unions, or governments. According to John Belliveau's *Little Louis and the Giant KC,* Irving backed N.B. premier Louis Robichaud's Liberals in 1960 to punish the Tories for flirting with a European pulp and paper company. Irving turned against Robichaud when the premier opened the province up to other industries and passed new laws to equalize

THE MYSTERIOUS EAST

taxes and social services throughout the province. The equal opportunity reform program, inspired in part by a "Saskatchewan mafia" of former CCF civil servants, ended Irving's tax shelters in isolated municipalities and introduced universal suffrage in muncipal elections, according to Della Stanley's *Louis Robichaud: A Decade in Power*. The Irving press whipped up anti-French hysteria, claiming the government planned to "rob Peter to pay Pierre." An Irving paper complained that Robichaud "seems to have forgotten what he owes Mr. Irving." Robichaud countered with the slogan: "If you want to run this province, then offer yourself to the people." That 1960 vote is considered the only provincial election that Irving ever lost.

As much as Canadians hate winter, there are advantages to life in the "true north strong and free." In Canada the absence of a sunbelt has sure paid off for labour.

The comfort zone in the Southern United States has been a haven for the worst working conditions, social legislation, labour laws, and rednecks in the western world. From the days of slavery on, employers have created a Third-World social climate and low-wage ghetto that undercut union strongholds elsewhere in the country.

In Canada there have been pressures to turn have-not provinces into a sunbelt of low-paid labour. Maritimers in particular have been victims of mass unemployment as well as of sharecropping methods in agriculture, fishing, and woodcutting. But the east coast has never become Canada's deep south. Part of the credit goes to universal social programs that maintain basic standards across the country. Part of the credit also goes to high rates of unionization and a series of union campaigns for national rates of pay.

K.C. Irving, New Brunswick's most powerful employer, built an industrial empire and social system on a form of double-entry bookkeeping that gave him the best of both worlds. He charged high regional prices and provided low regional rates of pay. In 1960, local rates of pay meant average yearly incomes of $1,853 in Kent County and $3,604 in Saint John, compared to $4,000 across the country.

Irving's ultra-modern Saint John refinery, with double the productivity of other plants on the continent according to a *Canadian Labour* report in March 1964, imposed the local wage formula. The pay and benefits package was a dollar an hour below refinery rates elsewhere. Irving blamed these low rates on federal government policies that discriminated against New Brunswick. In 1963, Irving's 150 refinery workers struck to win equal pay for equal work and to resist the pressures to turn the Maritimes into a non-union, low-wage ghetto.

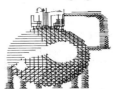

Roy Kippers moved to the Maritimes to escape Uppity Canadians. A Dutch immigrant who served time in Nazi concentration camps, he settled in Toronto in 1950 and found steady work at the B.A. refinery in Clarkson. But he couldn't stand the "holier-than-thou" atmosphere of "Toronto the Good," he says. So in 1959 he jumped at the chance for a job at Irving's new refinery in Saint John. He was looking for a change in his way of life, not a chance to get active in unions. He wasn't afraid of getting involved. "After being in a concentration camp, I fear no man," he says. But unions just weren't his thing.

Kippers was one of several experienced operators brought in for the start-up of the new refinery. They were all paid the national rate, but they were all expected to leave once they trained locals to take over. Neither Kippers nor Irving knew that some in the start-up crew came at the behest of the Oil, Chemical and Atomic Workers Union, to help organize from the inside. OCAW director Neil Reimer made a point of encouraging top-rate instrument technicians to transfer to new locations. The best of these, Bill Lexmond, enjoyed an international reputation for his technical skills. Reimer enjoyed Lexmond's ability to make sure plants didn't work out production kinks until a union was recognized.

In 1960 the refinery workers at Saint John voted 140 to 24 to join OCAW. A year later they got their first contract. Two years later they organized 12 marketing units across New Brunswick and Quebec. In 1963 they bargained for the same salaries as other oil workers across the country, a raise of $1.10 an hour. This dispute went to a government inquiry, which reported: "It is one of the major holdings of management that they do not wish to disturb local conditions regarding wage and fringe benefits" and "upset the applecart."

The inquiry noted that two other Maritime refineries paid national rates and proposed that Irving do the same within three years. In response, Irving volunteered a raise of 15 cents an hour. On September

11, 1963, refinery workers voted 132 to 4 in favour of a strike. The strike started five days later.

The strike quickly became a battle of wills. Irving, after all, was a one-man show. He answered to no corporate board for production losses, and his wealth left a lot of leeway for stubbornness. "When you threaten to hit him with a hammer, he says 'try it,'" Reimer says. For their part, workers had to screw up their courage to take on the giant company. "Irving wasn't a father figure in this province," Kippers says. "He was a god."

Donald Campbell was a lay minister who worked at the refinery. A recent convert to unionism, Campbell edited the union newspaper in which he charged: "This province is run and controlled by a mere handful of men who are striving for power and might with the filthy lucre, the desire for which the Bible condemns." In December 1963 *Union News* reached for another Biblical parallel, proclaiming the strike a "David and Goliath battle."

It felt that way. Apart from owning major chunks of the economy, Irving maintained excellent political connections. His support for the Liberals had helped them win the 1960 election. New Brunswick's top judge and Saint John's member of the provincial legislature, a cabinet minister, had both been on Irving's payroll. The Irving media provided little news coverage of the strike. Irving kept his automated plant running with a skeleton staff of supervisors, even putting in a shift himself.

Strikers tried to hit Irving in the pocket-book by boycotting Irving gas. The provincial court granted injunctions outlawing information pickets at gas stations. Saint John's Irving-owned radio station refused to run a union ad. The federal Board of Broadcast Governors agreed with Irving that this refusal didn't violate the station's obligation to neutral coverage. Another federal agency, Central Mortgage and Housing Corporation, refused to accept delays on mortgage payments from strikers. It took months before the union got that decision reversed and normal personal hardship standards applied.

The strikers knew what they were up against. It scared them. Some wore disguises and grew beards so they wouldn't be recognized on picket duty. But not one striker crossed the picket line. They "stuck together like glue," says Ron Duncan, who was OCAW director when the strike was on. "It almost brings tears to my eyes. They fought against odds that most people would have given up on. They established something in the Maritimes that just hadn't taken place before."

Eventually, strikers caught the spirit of revolt. "New Brunswick is very conservative in many ways," Roy Kippers says, "but they're good neighbours. If you're different, they let you be. There's a lot of freedom." Strikers and their wives went door to door to win support for the boycott. Fred Kahanek, on OCAW staff, was the union's stuntman. He led a delegation to city council to demand a monument honouring the courage of those who stood up to K.C. Irving. He took one

87

demonstration to the historic Loyalist graveyard opposite Irving headquarters and led a chant about how the earth was shaking from all the pioneers turning over in their graves because of Irving's practices. The climax was a march down Saint John's main street, where strikers cheered as Irving was hung in effigy. It was like protesters burning the U.S. flag at a demonstration in Washington.

The union made the strike a national *cause célèbre.* "Are you going to allow that a group of workers be slaughtered by an employer whose trade union notions are those of feudalism?" the local union president asked unions across the country. "We are on the battle line. It is the struggle of all the oil workers we are leading, and it is also the struggle for the emancipation of the working class of New Brunswick which we want to win." Kippers went on a national fundraising tour. He called his employer King Irving and likened the Irving empire to the Dutch East Indies Company, the grandfather of world monopolies. The appeal struck a chord in the West, *Union News* suggested, because Standard of California (a major shareholder in Irving's refinery until 1988) used its Irving connection to dump cheap oil in Canada at a time when Western oil couldn't find a market.

Donations flooded in. According to union records, over $106,000 came from OCAW locals across the continent. Other unions, from waiters in Sydney, Nova Scotia, to fruit and vegetable workers in British Columbia, donated $25,000.

In the Saint John area itself, the strike was endorsed by the police association, the firefighters' association, the ministerial association, and local Micmacs. Local labour groups flirted with a one-day sympathy strike to support the refinery workers.

Some of the stunts and rhetoric were too personal and vindictive for Reimer's liking. He had studied Irving carefully and had a hunch that the union's style would only harden Irving's resolve to humiliate his tormentors. Reimer thought it better to play to Irving's ego, reminisce about the old days of backbreaking labour when both worked the Western harvests, talk about the possibility of showing up Imperial — which Irving detested — by treating his employees better.

Kippers, who later turned in his refinery overalls to become a therapist for troubled teens, says the union stance worked on Irving's psychology. "He thought the strike would collapse. Then he saw his public image as a benefactor of the region going down. That's probably why he settled. He could've held on because he has oodles of money." (Despite repeated requests, Irving management officials refused to comment on this or any aspect of the strike.)

Settlement talks got under way in December, secretly helped by the deputy minister of labour. Irving insisted on settling personal scores. He wouldn't take back the man who had called him King Irving, he said. He offered to raise wages to national levels, as long as the raises weren't seen as a victory for the union. They would be based on personal merit increases totally under his control. Union negotiators were tempted to let Irving save face. Chief negotiator Fernand Daoust postponed an imminent Quebec strike of Irving truckers to keep tensions low. Daoust warned strikers in Saint John that this was probably Irving's last offer, that they might be voting themselves out of a job if they rejected it. But the strikers voted 79.8 per cent to reject -- "an indication of the determination of our members," Daoust wrote union headquarters.

On January 4, 1964, Daoust pulled the plug in Quebec, and 150 workers in 13 locations joined the strike against Irving. The formal issue in Quebec was Irving's demand that he be allowed to contract out work to part-timers. National bargaining rates and seniority rights were also at stake. But the Quebec workers had been primed to support the Saint John strikers since November and were itching for a fight. The Quebec strike scuttled further discussions with Irving in Saint John. Irving was also outraged by what he considered an insulting leaflet from Saint John refinery workers. All talks were postponed for two weeks.

In Quebec, Irving unleashed his fury. He quickly moved in strikebreakers and hired 150 private police to protect them — one for every striker. A battery of ten lawyers came forth with injunctions against all pickets. Within days 20 strikers were arrested. Civil action was taken against three strikers and the OCAW was sued for $625,000 in damages for lost business. On January 24 Irving insisted that his December offer was his last. On January 29 he issued a press release stating, "It is the prerogative of management to determine who is qualified for promotion." The union, he warned, "can hold out indefinitely by asking for something which management cannot agree to give."

The union leadership started to waiver. Daoust saw no end in sight. He had a hard time holding together the far-flung strike in Quebec. He didn't think the union could afford the expenses of Irving's legal bullying. Refinery strikes were drawn-out and expensive, and the international strike fund couldn't take the drain of all the strikes it faced that year. The chief counsel from OCAW's international headquarters, John Tadlock, came to New Brunswick to lean on local leaders to accept a settlement. The refinery workers stood up to Irving's threats, refusing to give him total control over "merit" increases. That forced government conciliators to pry a better offer from Irving in late February. By March 1 there was an agreement for a merit system geared to national rates and subject to the union grievance procedure. In effect, the increases would be automatic. Saint John refinery workers voted 88 to 16 in favour.

The Quebec strike settled the next day, with exceptions to the formerly airtight contracting-out clause. Irving dropped his $625,000 suit against the union for damages in lost business.

But he got his pound of flesh. The multi-millionaire's last condition for settlement was $2,000 to compensate for the damages done to his personal reputation. Union negotiators agreed, on condition that the deal was kept quiet. Perhaps they were too embittered to enjoy making a public spectacle of valuing Irving's reputation at $2,000.

On Friday, March 13, Daoust made delivery. "This payment was made in solid cash, so to avoid any use by this miser in Saint John of any photograph of a cheque that we could have given him," he wrote OCAW director Ron Duncan. "This was my black Friday."

WORKED OUT OF A JOB

Oil and chemicals have a greaseball image. So the high-tech story of the 1950s was the Avro Arrow. When John Diefenbaker ordered it scrapped he was accused of dealing a death-blow to advanced science-based industry. Whereas Canada's oil and chemical industry, on the leading edge of world technology, plodded along without space-age glamour.

Most of Canada's oil and chemical plants were state-of-the art, brand new in the 1940s and 1950s, with all the gadgetry that money could buy. Their owners spared no expense, ploughed back earnings into yet newer and faster equipment. No other industry came close. According to Douglas Fullerton and H.A. Hampson's 1957 study *Canada's Secondary Manufacturing Industry,*

oil refineries reinvested 60 per cent of their earnings between 1948 and 1955, six times the average for all industries. Petrochemicals took off in the 1950s and chemical companies invested $100 million a year into all branches of the industry, according to the 1957 Royal Commission on Economic Growth. In 1965 investment in petrochemicals shot up by 25 per cent, the *Financial Post* reported in June 1966.

The investments paid off. Oil and chemicals led the pack in every index of productivity. From 1950 to 1962 all of Canadian manufacturing boasted a 70 per cent productivity increase. The chemical industry weighed in at 120 per cent, according to George Eaton's 1968 federal task-force report on chemical bargaining.

Workers outdid themselves every year. In 1951 an oil

worker refined 4.7 barrels of oil an hour. In 1956 a worker finished off 11.1 barrels an hour, according to statistics in the April 1957 *Union News*. Chemical workers kept up the pace. In 1947 a chemical worker added $81,000 of value and in 1957 the same worker added $138,000, the Royal Commission noted. In the 1960s, output per oil worker jumped 184 per cent, according to Ministry of Trade and Commerce figures. In 1976 a union brief to the federal Anti-Inflation Board stated that oil workers — half a per cent of the workforce — created 16 per cent of all industrial profits in the country.

They worked themselves out of a job. In 1957 there were 8,000 refinery workers in Canada. By 1965 there were only 6,300. Constant expansion of new products kept chemical employment stable at the 50,000 mark, but the workforce at most plants shrank. Factories making acids hired 150 workers in 1947 and 142 in 1953, while factories making fertilizer went from 86 to 56 workers in the same period, the Royal Commission found.

Oil refineries didn't shrink, they shut down. Economies of scale were such that operating costs per barrel were slashed by five times when a plant upscaled from 1,500 to 70,000 barrels a day, Alberta's 1968 gas inquiry noted. A new and larger refinery would service a wider area and replace several smaller plants. Pipelines helped to nudge that trend along by removing the need for refineries to locate close to customers. Each new factory of the future produced several ghost towns of the past.

A TIGER IN THE TANK; A TIGER BY THE TAIL

Energy and chemical workers were off splitting atoms when most Canadians were out splitting wood. Wood was still a more common fuel than gas, the gas pools in Alberta's Turner Valley were burnt off as waste in nightly million-dollar light-shows, and cooking with gas was only a pipedream until the trans-Canada pipeline was built in the mid-1950s, William Kilbourne's history of the pipeline notes. By that time, Sarnia refiners were cranking up their first catalytic converter, cracking the carbon chain of petroleum, and controlling the spin-off of byproducts. "We were reaching the edges of technology at that time," says mechanical engineer George Rowbottom, who helped install the first catalytic converter. Automation in the "miracle industries" didn't wait for the bells and whistles of the hulking computers of the 1960s, the transistors and microchips of the 1970s, or the robots, gene-splicing, and superconductors of the 1980s. Many of these tools just caught up with the atom-splitting and instant information feedback long established in energy and chemical production. "Computers just gave an electronic picture of what the old digital control panels looked like," Rowbottom says.

Inventors played multisyllabic scrabble with molecular combinations that filled dumps with disposable garbage and threatened to dump workers from disposable jobs. Energy and chemical workers were among the first to confront the downside of high tech in the area of healthy and secure jobs. They had to learn to bargain as smart as they worked. In 1965, national bargaining gave them the imagination and discipline to pull off a knock-down drag-out strike that won precedents that are still the envy of many workers 25 years later.

When Neil Reimer worked at the Co-op Refinery in the 1940s, fine-tuned octane gas was for showoffs. "Skunk gas" was good enough for old Model Ts and low-compression motors. Refinery equipment gave off as many knocks as the gas. "If we could run our plant 60 days without shutting it down for repairs, we were breaking records," Reimer says. "Today, most plants run six years without a break."

The tools of Reimer's trade were one step ahead of the pioneer issue used to refine gunk from the continent's first oil rush in Petrolia, Ontario. In early days the Sarnia-area oil patch boasted 27 refineries built in the fashion of whisky stills, or like the sugar kettles used for making maple syrup. Workers heaved rocks at the boiling still, listened to the echo, and figured out the oil level.

Long before Reimer's time the vats had been sealed and the rocks replaced by gauges. Still, when Reimer began his work the oil oven was a sticky business. It got "coked up," and the sulphur in the crude corroded equipment. Periodically, production was shut down for a "turnaround," or maintenance. That was Reimer's job. He was the oil industry's chimney sweep. He hammered and chipped away at the coke, finishing each day black with soot. His wife says the smell of gas never left him.

Once the pots were boiling again, operators kept an eye on the oil broth through a bank of gauges that fed back figures. Instruments were made of low-quality metals. They weren't smart, and they needed constant repair. Operators worked up their own calculations every hour and adjusted the recipe as necessary. At the end of a shift they added up the day's tally and wrote it in a logbook.

By the early 1960s, long after Reimer left the Co-op in 1951, those rustic methods were history. Catalytic cracking (adding a catalyst to speed the break-up of the carbon chain) and hydrogen cracking made better gas and gave more control over byproducts. New methods required not only stronger equipment that withstood enormous pressure and heat but also tighter controls over hundreds of rattling carbon chains. Ingenious air-powered tubes, springs, and bellows sent signals to operators. In the 1950s, Petrofina's Montreal plant, for instance, had an automatic logger that analysed four hundred process variables and pinpointed problems. By the late 1960s small and hardy transistors let electricity take over from air pressure and opened the latch to the computerized factory of the future.

As union director, Reimer tracked this revamping of the industry to get a handle on bargaining and organizing trends. The good news was that workers deserved a raise. In a research report published in the April 1957 *Union News,* Reimer argued that wages, up 28 per cent since 1951, lagged far behind productivity, up 136 per cent, not to mention profits, which were up 67 per cent. At the same time, the days of organizing oil workers were numbered. The union must expand into other areas or die, he told OCAW's Canadian Council in 1961. If the union had stuck to its base plants, it would have lost two thousand members, he said.

Automation had also changed the power relations in strikes. Workers could set up picket lines, but strikebreaking was done electronically. A skeleton crew of supervisors bunked in for the duration and let the technology run itself. In an industry that relied on the mind, strikes were a battle of wills. For traditional union tactics, this was the oil shock of the 1950s and 1960s.

Apart from his sense of union strategy problems, Reimer's farmboy upbringing gave him a sixth sense about the social cost of new technology. As a child in the 1920s he saw horses replaced by tractors. As a teen in the Depression he watched horses haul tractors and cars, because farmers couldn't afford gas. Farmers went broke because their modern methods and technology produced such a glut that prices collapsed. As a young man Reimer joined the exodus of farm youth, driven off the land by a mechanical enclosure movement that demanded ever newer equipment and ever larger acreage to make ends meet.

At agriculture college in the 1940s, Reimer was taught by Grant MacEwan, one of the West's most popular writers and a future Lieutenant-Governor of Alberta. MacEwan was an early animal rights crusader who mourned the passing of the old-style farm where people followed rhythms that linked them with nature. He warned that energy companies would jack up their prices as soon as horses were traded in for heavy metal. "He got me thinking that the way we were going wasn't going to solve farmers' problems," Reimer says.

Reimer brought this horse-sense to his union job, holding onto this gut instinct that progress could bring poverty, not wealth. To make sure workers and communities got their share of the wealth they created, the absolute rights of property had to be challenged. Reimer's 1957 research report reflected this focus on first principles. The wealth created by increased productivity belonged to workers and the community, not just owners, he wrote. "Certainly this is the preferred way to benefit our people — rather than have the wealth from our national resources drained off in the form of profit to foreign corporations."

In 1960 Reimer heard "Shorty" Phelps, the International's one-man think-tank, at a brainstorming session for union district directors. Phelps said workers, who invested their lives in a company, had rights on par with shareholders, who only invested money. The company owed them a return on their investment if a plant closed its doors. All the directors liked the idea. It expressed a radical union demand for human rights in the conservative language of property rights. But when the time came to pick a volunteer to pilot the concept in bargaining, they talked about Canada as the only district that could make it work. "Bravery is often akin to ignorance," Reimer says, "so I said I'd try it."

Staff coined the term "reimer reason" to ridicule this pet project for tilting at petrochemical windmills. It was heresy, a direct attack on the sacred notion of property rights. Routine union bargaining didn't normally confront those rights, at least not head-on. But Reimer thought automation put the meaning of property rights into sharp relief. Property rights gave employers the power to treat workers like any other disposables of a throw-away society. He wanted to take on that power.

The union skeptics were wrong. Oil workers were ready for brave new thoughts. They had a way of looking at their jobs and thinking about the world that made them receptive to new approaches to job security.

Oil workers tended to see their jobs as careers. Career security was on their minds when they joined the union. Ray Jamha saw that right away. He knew he couldn't make the same pitch to oil workers that he used to make to meatpackers in the blood and gore of the kill room. Wages were too good, and supervision was too lax. When he talked about career security, oil workers signed up.

With automation, prospects for career-minded workers became discouraging. Suddenly all the challenging jobs seemed to go to outsiders, to either contractors or university grads. Contracting out — commonly defined as an issue of the lean and mean 1980s — grew like a weed in the energy and chemical industry of the 1950s. Managers streamlined their workforce by putting regular staff in charge of basics. Contractors got to handle all-round maintenance during turnarounds, and they did the jobs that needed a creative twist. In October 1953 the *International Oil Worker* called this trend "a major cause of the present feeling of uncertainty among oil company employees." They saw their chance at broader training walking out the door each time a contractor left.

High-tech systems were also hard on the ego. North American managers rarely used technical breakthroughs to promote "soft-path development," either with the physical environment outside the plant or with the social environment inside. George Rowbottom was a design engineer at Canadian Oil's Sarnia plant in 1962, a year before it was taken over by Shell. The takeover coincided with a switch to a new central-control system that replaced four distinct control houses, he says.

Old-style operators treated their control house as a home away from home. They installed hot-plates and other comforts of home. And a controller's house was his castle. "In the old days, managers almost had to have permission to go into a guy's house," Rowbottom says. "He was the king." The centralized system evicted them. "There was apprehension from then on as to when the next shoe would drop," Rowbottom says. "I have the greatest respect for the qualities of a good operator. These plants are like a bunch of spaghetti with lines going hither and thither. The ability of these people to diagnose from instrument read-outs was just like a doctor who figures something out by thumping and listening. There's a psychological change when you take people like that and downgrade their standing."

Oil workers had time on their hands to nurse these wounds and talk about solutions. They were also as close to a leisure class as workers got. A good supervisor breathed easy when he saw workers with their feet up on a desk. That meant there were no problems and the plant was making money. So operators had a leg up when it came to reading and debating. They were a bit like the old-time printers who became early labour movement leaders simply because their jobs gave them time to read and think.

Stu Sullivan, who started at the Moose Jaw B.A. plant in 1956, was a graduate of this factory night-school. "We read, we discussed, we debated. Because in an average shift, you would have at least five hours to do that — then, if something went wrong, you'd earn your salary in two hours," he says. "The topics of conversation used to be the union, the industry, sex, and the boss." New technology often cropped up as well.

Unlike the old-time printers, however, oil workers were strictly modern. They had no ancient rules of thumb or rituals that dictated how a job should always be done. Because the industry had no unions until the 1940s, there were no jealously guarded craft-union rules around particular jobs. The "featherbedding" or "make-work" rules common to historic groups such as longshoremen, printers, railway running trades, and building tradesmen were simply not considered. Holding onto the past wasn't an option for the industry, the union, or the workers.

On their own, without much union prompting, oil workers outlined bargaining demands around job security that focused on sharing the benefits of, not stopping, new technology. In 1962 they filled out union questionnaires on bargaining priorities. "Job security is the biggest common denominator," OCAW's Canadian Council executive reported. Workers ranked job protection, the shorter work week, early pensions, and longer vacations ahead of wages. In January 1963, *Union News* reported that the national oil bargaining conference highlighted lifetime job guarantees for those with five years' seniority "to protect workers against the spreading layoffs brought on by automation."

Unique in its time, the demand reflected Reimer's philosophy, oil workers' culture, and national bargaining's forward-thinking bent. Lifetime employment guarantees are widely recognized as key to the Japanese success with high technology. But the industry gave a flat-out no when the issue was tabled in 1963. The union's strike fund was exhausted, so the bargaining team backed off. The demand stayed on the union's books for the 1965 bargaining round. It was on the union's wish-list tabled in November 1964. It was on the short-list tabled just before the strike in September 1965. But it wasn't put front and centre until the strike began. Which is a story in itself.

There was never any doubt that a long strike was brewing in 1965.

Ron Duncan, who replaced Reimer as union director in 1963, set the mood for a do-or die battle. "We must take the position that we would rather go down to defeat in a real fight than relegate ourselves to the position of beggars picking up whatever crumbs the oil industry wants to throw us," he wrote local leaders in March 1965. In May he set up local strike committees that masterminded a nation-wide shop-floor campaign to buttonhole managers. On June 18 oil workers sported glaring orange buttons featuring the mystery letter "P." It stood for progress, for parity, for "piss on you," supervisors were told. It was an excuse for pranks, pandemonium, provocation, power to the imagination. Members "really went to town, sticking them on the supervisors' asses and all the machines and trucks," Duncan says. Duncan stepped up the psychological warfare in July. "Just bug the hell out of them," he told local leaders, advising them to work to rule so management would get "an indication that their operations can be slowed up without direct strike action." Ray Davidson, a public relations expert with the International, gave lessons from the U.S. civil rights movement on disorderly conduct that would fall just short of outright disobedience. Frazzle their nerves by staging far-out, attention-grabbing incidents, he wrote in one confidential memo. He plugged the "reverse walk-out" — circling supervisors at quit time to harass them about bargaining and to block off the next shift.

Bargaining minutes at Shell's Manitoba plant cite "a five-minute flurry" by greenhorn staff rep Reg Basken, who told the company "that the days of watching Imperial Oil offers were over and instead of watching Imperial Oil, it was time to listen to the demands of the Oil, Chemical and Atomic Workers." Basken recalls that he was cut short by management's negotiator, "a Texan who drawled that `we can pay a hundred dollars an hour. I'm proud to say we never plead poverty, but we only pay what you squeeze out of us. When does the squeezing start?'"

Management was also geared for a showdown. In October 1964, *Executive* magazine said OCAW's organizing scorecard "reveals some cause for management alarm." The union represented 40 per cent of the industry but had drafted 21 company unions behind its bargaining drive, the magazine warned. Company bargainers dragged out talks, refused all offers, showing they weren't going to be pushed around.

In a battle of nerves, plants became theatres for cat and mouse games of provocation-escalation. Managers would try to show who was boss. They started stockpiling, building up reserves for a long strike, a reminder of who held control over the purse strings. Workers refused overtime. Management pushed. When a Clarkson lab worker was ordered to work overtime he turned it down and got suspended. His workmates walked out in solidarity.

On July 2, Shell management in Vancouver issued emergency overtime orders. Workers refused, saying it was "awfully brazen" to ask for co-operation when none was offered in negotiations. Shell didn't ask for co-operation. It issued ten men strict orders to report for overtime or risk suspension. When it came time to change shifts, workers staged a sit-in until management agreed to withdraw the threats. Shell threatened to fire local union president Jerry Lebourdais unless the sit-down stopped.

Workers took that lying down: They had a sleep-in. Picket lines went up. The next shift refused to cross. There was scuffling between workers and supervisors.

The next day Shell produced an injunction. It brought in the RCMP to evict the hundred or so "trespassers." The strikers held their own musical ride, falling in behind a bagpiper and marching out in columns of four to the cheers of friends and families gathered around the gate. The next day Shell dropped charges against everyone but Lebourdais, the revolutionary *enfant terrible* of one of British Columbia's first families and a constant thorn in management's side. His case went to arbitration, which later upheld the firing.

It took cool judgement and stern discipline to hold back from premature confrontations that would dissipate energy and resources and undermine nation-wide co-ordination. The strike deadline came just in time. On September 13, a phased nation-wide walkout of a thousand refinery workers began. B.A. was targeted across the country. Shell was targeted in Vancouver and Winnipeg when it offered to refine B.A.'s crude there.

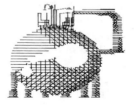

From the start it was clear that automation would define the outcome of the strike. Not just as an issue to be settled, but as a new set of rules for conducting a strike.

Old-style strikes work wonders on an assembly line or construction site. The most important factor — that it's workers who make the wheels turn — comes right to the surface. Not so in high-tech industries that run on momentum. Once supervisors are brought in to work — as they were at B.A. in Clarkson and Moose Jaw, and at Shell in Winnipeg — a strike becomes a waiting game: waiting for a major breakdown that supervisors can't handle; waiting for work to pile up; waiting for workers' savings to dry up.

That's the main contradiction of a strike based on the issue of automation: Workers strike to get rights on par with machines, but the machines keep working. For management, the secret of winning is to keep workers out of sight and out of mind. For unions, it's the opposite. The strike has to become a public issue that won't go away.

The road to winning public support led back to Saskatoon. Neil Reimer, a staff rep at the time, was assigned to the B.A. strike there. That's where he threw a wrench into the company's, and his own union's, game-plan.

Reimer started by asking for an extra day to consider management's latest offer, knowing that the company had already made the complicated arrangements for a shutdown. It was too late for Saskatoon management to backtrack. When workers reported at the plant the next day they found themselves locked out, with security guards and guard dogs at the gates. A guard obliged photographers by demonstrating what a doberman could do if it lunged at a striker's throat. The bared-teeth picture made front pages across the country. The next day the Saskatoon *Star Phoenix* reported that the managers had removed the dogs, but it was too late. The union had drawn first blood in the battle for public sympathy, showing that management was going for the union's jugular.

Reimer may have upstaged the company in the national media but the show didn't play in Saskatoon, or in Moose Jaw either, where B.A.'s other plant was shut down. While sipping coffee in a Saskatoon restaurant, Reimer overheard some farmers in town for the day. They were saying that workers had nothing to complain about — their wages were already high enough. That's when Reimer realized the need to highlight non-wage issues. He decided to put front and centre the union's demand for protection from the layoffs created by new technology. "We almost had to do it to maintain local support," Reimer says. "I said the strike was to preserve our jobs and communities because of the wages paid there. I put the new technology issue in the mill once the strike was on. Then I called the union staff in Denver and Vancouver to get them to raise the issue's profile. That caught the imagination of the whole labour movement. Automation became the issue."

It sent management spinning. Their reading of union priorities was that money was the central issue, and that job protection had been thrown in as a wishbone to excite the ranks. Still befuddled by the union's shift of ground 20 years later, Norm Devos, a former B.A. personnel manager in British Columbia, told Reimer: "I've heard it said that it was something that you suggested and eased in around the back, as another talking issue to get off the point." Ted Gaunt, who was B.A.'s industrial relations chief, told Reimer: "There were certain things you asked for. And then your demands were something else. I hate to say this but I think it was a pretty good issue to pick up on."

For his part, Reimer is still mildly sheepish about having added this issue on his own. "I was a little nervous, because if somebody did that to me, they'd have likely caught a lot of shit."

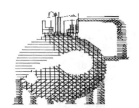

On September 18, 1965, after less than a week on the picket line, a union leaflet at Clarkson referred to the strike as a "total economic war." Total wars strike at civilians, not just troops.

To hit the company in the pocket-book the strike had to undercut its sales, not just production. So the union immediately launched a national boycott of B.A. products and appealed to the car-driving public: "B.A. unfair. Buy elsewhere." Picketing of gas stations in union towns across the country began on October 2. The boycott worked in Saskatoon, mainly because strikers let it be known that they'd mixed in additives that made motors seize up. In British Columbia it worked because people there respect picket lines.

But neither the strike nor the boycott hit Gulf where it hurt. The head office applied no pressure to settle, Ted Gaunt says. "As long as sales were okay, we were going along quite efficiently. There were no trucks being knocked over or burned. Nothing like that was happening."

Supervisors filled in to keep strategic plants working. Within a week they were pumping out gas at 74 per cent of capacity, a company bulletin of September 23 stated. Special newsletters, visitations, and taped messages from senior executives, lots of steaks, movies, and popcorn — not to mention triple pay -- kept the morale of the strikebreakers high.

In Ontario management felt so smug about the smooth functioning of the ultra-modern Clarkson plant that it talked about ordering a speed-up after the strike. "What are we fighting for?" a company news release asked on September 28. "It becomes increasingly apparent as the strike progresses that the Company is fighting for an improvement in the efficiency of operations and a discontinuance of the overlap in individual jobs."

In British Columbia union pickets hit their heads against a wall of corporate and government togetherness. The boycott had to be broadened because companies that weren't struck sold B.A.'s gas through their outlets. But when pickets surrounded the Imperial Oil terminal, the government slapped an injunction on them because the terminal was owned under a separate name. Pickets were banned from B.A.'s Vancouver terminal because it was on CNR grounds. Strikers couldn't picket Royalite, even though it was owned 98 per cent by B.A. And no one could picket gas stations. Collusion among companies was fattening; against companies, it was illegal.

The injunctions were put to good use. "We used them to decorate the office as wallpaper," Buck Philp, former B.C. strike co-ordinator, says. But the failure of spot picketing left the union with no choice. "We decided we had to shut down the whole west coast," Duncan says. Notice was served on November 15, 1965.

OCAW had a lot of cards to call in. In 1964 a group of "mystery pickets" marshalled by Reimer had shut down a Standard of California refinery that bought cans from a company being struck by the Steelworkers. Standard charged Reimer with conspiracy, which carried a six-year jail sentence. Reimer went into hiding while police scoured the town looking for him. In the evenings he signed himself into hotels under a false name. He passed away some days sitting in the courthouse, right under the nose of the judge waiting to try him. By the time he turned himself in, the strike had been won. It was one of the actions that defined the B.C. Federation of Labour's tough policy against touching any "hot goods" from struck plants.

Past favours aside, the entire B.C. labour movement adopted the B.A. strike as its own and put itself on a state of alert. On September 27, the B.C Federation of Labour had backed the B.A. boycott because, the federation said, it was "the first real confrontation in British Columbia on the key issue of automation." By October unions were refusing to allow contact with "hot" oil. Construction on a major power project was stopped until a barrel of B.A. oil was removed. OCAW hired a plane and a boat — the "African Queen" — to picket tugboats and logging ships. Crews refused to unload ships fuelled by scab oil.

On November 2, the B.C. Federation of Labour adjourned its convention to picket B.A. offices in Vancouver. The Fed's support committee included heads of 30 unions. On November 15, the Fed hosted a meeting of all unions in the province, including those outside the Canadian Labour Congress. The three hundred leaders there asked the oil workers to postpone their total boycott until November 24 so they'd have time to build a strike that could shut down the whole province. Organizers drove, flew, and boated to remote parts of the province to win support for a likely showdown. There were 150,000 people ready to take part in illegal walkouts. Teamsters said they'd park their trucks wherever their tanks ran dry. The government's hand was forced. Labour minister Leslie Peterson personally took over conciliation of the bargaining teams. On the morning of the strike deadline Premier W.A.C. Bennett called a press conference. Union and management leaders sat poker-faced -- both groups had never bargained in a fishbowl before. Both did not know what to expect from the autocratic and unpredictable "Wacky" Bennett. Bennett outlined a settlement that he hoped would bring the province back from the brink. The union would have wage increases of 35 cents an hour and protection from layoffs. Duncan, looking sullen and downcast, left to consult with his bargaining team, and an hour and a half later he offered reluctant support. Ray Davidson of the International says Duncan missed his calling as an actor. The oil companies had to match his show of compromise later in the day. With 12 hours to go before general strike deadline, the strike was called off.

The oil companies lost their "unilateral right to throw people on the scrapheap wherever, whenever, and however they wanted to," *Union News* stated. Each plant got a joint union-management committee on automation. If the committee got into a deadlock, the company couldn't lay workers off until the next round of bargaining. Management had to give six months notice of any layoff resulting from a change in technology. Those laid off got a week's severance pay for every year worked. The company had to co-operate with the government and union to retrain workers for new jobs.

Some seven hundred striking oil workers in British Columbia had greased the skids for a national settlement in their own industry and similar settlements in other industries. They couldn't have done it without help from their friends. Some fumed at the strike, saying that never had so few caused so much trouble for so many. There was another way of looking at it, according to the December 1965 *Union News:* Never have so many done so much to help so few. "This display of labour unity posed the most drastic threat to industrial peace since the fabulous Winnipeg General Strike of 1919," the paper said.

That same December the industry magazine, *Canadian Petroleum,* carried a sour grapes editorial on this theme. It complained that OCAW had ganged up with the B.C. Fed to use labour's political clout to turn negotiations "into a rubber stamp affair." The government refused "to protect its citizens from the violent power-mongering of organized labour." As a result, the magazine said, "The unions have gained provincial acceptance of automation as an issue. For the first time, notification of layoffs and severance pay (previously company policy) will be written into contracts." More galling still, the union victory had been snatched from the jaws of defeat. B.A. had felt so strong it had prepared a final offer that was lower than its original one.

OCAW rated its new job security clause as a "significant foot in the door." In hindsight, that judgement was too hopeful. In the years since, without the same kind of solidarity mobilized by the oil workers in 1965, few unions have been able to pry the door open further.

Strikes are tough, so tough that they bring out the soft side in everyone.

A cornball story in a Clarkson leaflet got reprinted in every local strike update. It concerned two oil workers who always horsed around together. When one moved on, the other explained why he didn't miss him. He said his buddy was only good for laughs, that they never cried together. The story carried a moral: "Friends are friends if they are willing to expose themselves to sorrows that are personally shared." And true brotherhood comes from someone who "eases your pain because he knows what hurts and where." Macho men from the 1960s.

For all the hardship, many workers cherish the intimacy and community of a strike. They savour the common cause, the pranks, the chance to get anger out of their systems, the feeling that they can make a difference. They enjoy the free time, new friends, and rediscovery of simple joys. They're not always keen on trading that in for the old shift schedule. The end of a strike is often a tug of war between the mortgage on the house and the mortgage on the heart.

These kind of feelings caused a few hitches in extending the B.C. settlement across the country. It carried more slowly than the union expected.

In British Columbia, B.A. workers stayed out until Shell and Imperial workers got the same deal. One member, who owned a country lot, invited all strikers who were still in debt to come and chop down free Christmas trees.

In St. Boniface, Manitoba, Shell workers didn't go on strike. They were fired one at a time when they refused to load B.A. supplies. They stayed out until mid-December when everyone was rehired. Hank Gauthier, who was local president and later became an officer of the national union, says it wasn't just a sympathy strike. "We were the same blood. When we organized, we got the same blood transfusion. That multinational just wasn't going to get the best of us. Once you taste that, you can just bulldoze. It's just a matter of time, and knowing how big a bite to take."

A high point at Clarkson was the cloak and dagger game strikers played with a professional strikebreaker and spy, Hugh Brian Gallagher. According to Marc Zwelling's The Strikebreakers, Gallagher worked at various times for the RCMP and B.A.'s credit department. He may have been behind an offer of "help" that local union leader John Kane received from a strikebreaking supervisor. If Kane would bring in the dynamite, the supervisor promised he would blow the plant up. Kane refused. Gallagher was, more definitely, behind the bugging of the local union office. When the operation was uncovered the unionists informed the police, who did nothing. So the union invited Patrick Watson and his CBC-TV crew from This Hour Has Seven Days to film the office staff ripping out the walls and laying bare the bugging equipment. Port Credit police took the evidence, and lost it. No charges were ever laid.

Clarkson workers stalled the settlement until they made their New Year's resolutions. "There is no need to make a hurry-up settlement just to change places with the scabs so they can spend Christmas at home with their families," their newsletter said.

Duncan walked the picket line in Clarkson "with a guy who had six months left to go to pension, with tears in his eyes. He was so happy that he had a chance to kick that goddam company in the ass. He'd have stayed out there, it wouldn't have mattered if they'd hauled him into the courts with every injunction in the books, he'd have been out there marching," Duncan says.

The original decision to strike at Moose Jaw carried by one vote. By the end, the strike was solid. Settlement talks bogged down until the dead of winter. One of the picketers, Stu Sullivan, remembers "a cold bitter night" when it was "35 below zero." He says, "All of a sudden someone knocked on the door of the trailer where pickets warmed themselves and said `take a look outside.' And there was yellow smoke coming out of cat tower. We heard the horns blowing and knew they had a cat reversal. We could hear safety valves popping and we knew they must be frantic. Everyone laughed and said: `We know one thing. The strike's going to be over in a week.' Couple of days later, it settled."

Saskatoon was the last to settle. Charles Hay, the head of B.A., had owned the plant before B.A. bought it. Duncan tweeked his nose by starting the strike there and Hay's nose stayed out of joint. Management demanded local concessions. A one-day boycott of local gas stations kept away all but four customers and got discussions back on track. Acceptance of the contract had to be shared with the women's auxiliary that local president Mike Germann organized. The wives packed the union hall every Saturday during the strike. "We kept them abreast of everything. We didn't hide a thing," Germann says. When the plant voted, the strike was over.

There was a lot of beginners' luck in the oil workers' victory.

Working in a new industry, they had no old technological axes to grind, no established craft unions that fought over jurisdiction rather than workers' rights. Industrial unionism, together with national bargaining, allowed them to exercize a bird's-eye-view of their problems. Though they belonged to an international union, they'd joined a young one with few entrenched leaders and practices, and with little inclination to interfere in Canadian affairs. U.S. staff offered advice, but never gave orders.

Never having fought a strike on this scale before, they had no memories to live up to, and nothing to unlearn. They felt free to try bold tactics like boycotts and general strikes that wiser --or is it more wizened? — leaders would have blocked, knowing the extreme penalties of failure.

Serendipity, structure, and strategy all played a part in the victory of one of Canada's smallest unions against the country's most powerful corporations. In the 25 years following, there were at least 14 shutdowns of refineries across Canada. The 1965 strike still pays dividends to those workers who have opted for transfers, retraining, or severance pay.

Looking back, Reimer thinks the strike was as much about philosophy as practicality. "Today, everybody wonders why we had the strike. But the strike was about ideology. `What do those workers know about technology?' the engineers asked. `We're the engineers. We thought of it.' The workers were being disregarded in every respect," he says.

In a backhanded way, the industry owned up to similar conclusions once the strike was over. The union's job-security demands caused "increased blood pressure among many oil executives," John McNulty wrote in a post-strike article in *Management* magazine. But many executives learned that they had to pay more attention to the human side of technical change, he wrote.

Employers have learned a few things since the 1960s. They've renamed automation "high tech." The new term conjures up images of progress, not layoffs. Reimer is worried that unions haven't kept up quite as well. He still sees the future of high tech from his mind's perch in the back-40 of a family farm. "In my lifetime, I saw agriculture go from 51 to 4 per cent of the workforce. There are no farmers more productive than ours, and they're making a product that's essential. But they're up to their eyeballs in debt. Productivity hasn't solved their problems.

"I can't visualize a solution under capitalism. I can only see a socialist solution. I suppose if I knew the answers I'd be in great demand as a public speaker. Or maybe I have the answers, and no one wants to hear them."

A STUDY IN CONTRASTS

OCAW's success with the issue of automation stands in striking contrast to the technical knockouts suffered by two of Canada's oldest and strongest union bodies.

In 1964 the International Typographical Union, Canada's premier union for over a century, was licked by automation. Until then, the "mighty typos" ran the printshops. Bosses weren't allowed in the room and had to put up strict workrules that verged on make-work or "featherbedding", much of it downright silly. Type had to be reset every time an ad ran, for example. The printers' ace was their skill — but it was also their only card. In the late-1950s, employers trumped it with "cold type," which reduced typesetting to typing.

Toronto's dailies demanded changes in 1964, and printers walked out in a huff. Except for the Mailers, other unions stayed on the job. Union pressmen and engravers took the places of the printers. For decades these craft unions fought over who got dues from what workers, and forgot that unions stood for workers against bosses. Though the strike dragged on for years, its defeat was sealed within days.

Sally Zerker's *The Rise and Fall of the Toronto Typographical Union, 1832-1972* provides a checklist of union errors. Though automation was taking place at newspapers across the continent, printers gambled all on one strike in one city. Members, kept in the dark, learned the issues from management memos on the eve of the strike. U.S. union leaders called all the shots. They didn't try to win support from other unions.

The printers at least fought. Railway unions were asleep at the switch. In the 1960s, when it was too late, railway workers blew off steam with rearguard wildcat strikes that seemed aimed at modernization itself.

Rosemary Speir's 1975 Ph.D. thesis chronicles the fiasco of whistlestop modernization in the 1950s and 1960s, when trains faced stiff competition from cars, trucks, planes, and seaways. Diesel fuel eliminated railway firemen, who once shovelled 28 tons of coal every 12-hour shift to keep the engines stoked. The use of diesel engines meant longer runs, heavier loads, and less stops for repairs and refuelling. Soon after, containerization lowered the boom on freight-handlers.

In a struggle requiring co-ordination and foresight, rail workers were divided into 18 unions, 17 of them craft unions run from the United States. Each union focused on its own jurisdiction, not the needs of all rail workers. A joint committee of rail unions didn't raise tech-change issues until 1961. "Somehow, we just didn't get around to it," one leader admitted. "For decades, the style was to bang the table, and then go out and have a drink with management," union lawyer David Lewis complained. After fifty thousand layoffs in the railway industry in 25 years, railway workers made limited gains from their wildcat strikes in 1971 — like closing the barn-door after the horse is out.

BUDDY, CAN YOU PARADIGM?

Columbus wasn't the first person to see things that didn't square with the idea of a flat earth. Galileo was no better stargazer than those who figured the sun rotated around the earth. Darwin wasn't the first to find fossils that looked more than six days older than Adam and Eve. All of them weren't heretics because of what they saw. They punched society in the inner eye, said that seeing had to be more than believing, made connections that were "outasight." That's what science historian Thomas Kuhn calls a "paradigm shift."

The 1969 strike of oil workers dramatized the paradigm shift, then just getting underway, in labour thinking about health and safety. Health and safety became a big labour issue in the 1970s and 1980s — to the point where everyone now talks about health and safety in one breath. Until 1969 they were two separate words, two separate issues, dealt with in two separate worlds. Safety was the responsibility of employers and ministries of labour. Health was the responsibility of doctors and ministries of health.

At the onset of oil bargaining, according to a report in the December 1968 *Union News*, workers demanded joint union-management health and safety committees to investigate toxic chemicals used at work. Their sense of health and safety as a compound complex term upset the entire safety establishment that had been in place since World War I.

In the 1880s workers had an old-fashioned, instinctive sense that health and safety went together.

Union briefs to a royal commission on labour relations made as much fuss about poor air quality and long hours confined indoors as they did about guards on machines. The great international workplace scandal of the era was "phossie jaw," a disease that ate away the jaws of young working girls who placed phosphorus on matches.

Health and safety issues were uncoupled around 1914 as workmen's compensation laws were passed. It wasn't an oversight when politicians called it workmen's compensation. They were thinking about "accidents" — falls, cave-ins, and cuts — that happened to men in mines, factories, and construction sites. The new regime was politically safe. Accidents were caused by "careless" workers who didn't follow strict management rules.

THE LONG HOT SUMMER

There was little emphasis on the injuries caused by chemicals and the workplace practices that afflicted the most careful workers. That was the bosses' business. At best, workmen's compensation picked up the tab after the harm was done. Boards never thought of interfering in the management right to run workplaces as the companies saw fit.

Health and safety, by contrast, was politically dangerous. The very act of linking the two issues shifted blame from careless workers to irresponsible managers, shifted expertise from professionals to activists armed with undoctored information, and shifted power from management's say-so to a joint say by the union.

Neil Reimer mumbled the words quietly, almost sadly. "Well, I guess we'll have to strike you." Ted Gaunt, chief negotiator for Gulf Oil and a trendsetter for labour relations in the industry, expected a bit more of a tantrum, at least some cussing and pounding of the table.

It wasn't much of a climax to the months of failing to agree on a starting point for the 1969 negotiations. The union had held firm for a national settlement. The industry had stood by a previous insistence from Imperial Oil on five regional settlements. Neither side budged. Until this critical issue was resolved, there was no point in getting down to specifics.

The real climax came later, after a five-month *tour de force*. By the time this lengthy main event was over, an entire province had gone through the industrial heaves, an election was thrown off balance, a striker was killed, and a new labour law had been left for dead. As if this wasn't enough, organized labour had been split into hostile factions, and two new bargaining issues for the entire labour movement had been placed on the agenda.

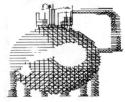

Reimer and the national office called the
strike to begin on May 23, 1969, in British
Columbia. "We felt that's where we had
the strength, pure and simple, where the economy was strongest and
wages highest," Reimer says. Members wanted to "do this together," he
says, so B.C. workers at Imperial, Shell, Gulf, Texaco, and Home all
walked out together. Only Union 76 and Phillips 66 were left in peace,
so consumers could buy union-made fuel. It was the first time in
memory that the Phillips 66 plant in isolated Fort St. John didn't have a
wildcat strike at least once a month.

At first strikers' spirits were high. They wore shorts, played catch
with frisbees, sunbathed. Strike co-ordinator Buck Philp stopped
shaving, said he wanted to look like a guerrilla fighter. He got nicknamed
Ho Chi Minh.

Before long, *commandante* Philp found himself in a war on four
fronts. The struck companies bunked in supervisors and kept up
production. The Employers' Council of British Columbia and the
Vancouver Board of Trade insisted on accepting deliveries from struck
plants. "It involves the basic principle of management's rights," said
F.G. Preskett of the Employers' Council. The government outlawed all
pickets that pressed a boycott of "hot fuel." The powerful B.C. Federation
of Labour refused to boycott any company but Imperial.

After their experience in 1965, oil workers knew they had to rely
on consumer solidarity and industrial havoc to bring the oil companies
to the bargaining table. In low-tech strikes, so the saying goes, the longer
the line, the shorter the strike. In high-tech strikes, the line has to be
wide. The union immediately branded all fuel from struck brands as
hot.

The Fed's secretary-treasurer, Ray Haines, wasn't keen to replay
the 1965 showdown. Haines came out of the retail industry and knew
the importance of boycotts. He developed the Fed's hot goods policy,
the strongest in the country. But he didn't want to push the button again.
In 1968 he had told OCAW's Canadian Council: "When the dust cleared
in that [1965] dispute and when we got back to normal, we recognized
that labour had, in fact, pulled an atom bomb. We really don't know
what to do in British Columbia for an encore now."

On May 29, 1969, Haines said a full boycott "would bring industry
to a standstill and adversely affect many." The next day, Reimer
denounced him for meddling in the business of an autonomous union.
A week later, an angry striker picketed the Fed's headquarters, charging
"B.C. Fed supports scabs." On June 18, OCAW's Walter Bissky pleaded
with the Vancouver labour council to join a general boycott. "We're not
dealing with one employer, we're dealing with an industry," he said.
His motion was defeated two to one.

On June 28, Reimer tried to patch matters up by co-signing a Fed appeal to single out Imperial for boycott. It sounded good. But Imperial was OCAW's Achilles' heel. The union only represented Imperial's Vancouver and Calgary plants. At any rate the preparation for a genuine consumer boycott was limited. The Fed's efforts had only included withdrawing Imperial credit cards from its own staff.

In the meantime, strikers couldn't hold their picket lines. Trucks were reckless when they ran the gauntlet, and several strikers were knocked over. A picket captain told the press that drivers "keep their toe down when they hit the line. They're really barreling out. Our guys have to be pretty spry to keep from being injured."

Two strikers sought revenge by tossing a chain over the electrical wires to the Gulf plant, destroying $250,000 worth of production when the lights went out. Slightly less illegal were union "flying squads" that picketed construction projects, planes, and steamboats using hot fuel. The courts immediately issued injunctions ordering the pickets to be on their way. The union collected 22 injunctions.

As the courts saw it, injunctions protected innocent bystanders who weren't directly involved and shouldn't have to suffer. By this logic, the courts forbad picketing against Imperial Oil Enterprises, claiming it was unrelated to Imperial Oil Ltd. As strikers saw it, injunctions let the oil companies get away with sales of strikebound fuel by using rented trucks and false-front subsidiaries. "The result can be a shell game of shunting responsibility from subsidiary to subsidiary," a frustrated picketer complained.

By July both the strike and the boycott were running on empty. "The companies feel the stamina of working supervisors will stand up longer than the pocketbooks of the union workers," the *Vancouver Sun* reported on July 2. The workers' "big weapon, the picket line, has been reduced to a token tactic," the *Sun* noted on July 16. The union settled in for a long strike. Half the strikers took on odd jobs so they wouldn't eat up strike funds.

By mid-August this "out on strike, out of mind" syndrome had driven Philp to desperation. He announced on August 13: "The strike has been pushed into the background. We have to make a change of policy. We have to show the public there is a strike on." The oil workers began a picket of ferries and city buses, causing one of the worst traffic jams in the city's history. Philp always checked with local union leaders before his crews went to shut a place down, but he forgot to check with the weather office before giving the program its official launch. The traffic jam left thirty thousand people stranded in heavy rain.

Philp called on the companies to return to the bargaining table, where both sides could choose a mediator to help them work through their differences. In the same breath he threatened a hit and run campaign against any large companies using petroleum products. "We know that whoever we picket will get an injunction against us whether we are in court or not, so we will wait for the injunction to be served and then picket someone else," he said. He was as good as his word. Workers at train stations, hydro stations, construction projects, steamships and the wheat pool stayed away from the job, until police chased Philp's crew away.

Philp avoided the Fed as often as he sidestepped the law. For the Fed, this was the worst time possible to defy law and order. There was to be an election on August 26, and a messy strike would make the NDP look bad. *Vancouver Sun* columnist Jes Odam wrote in mid-August that oil workers had become "political pawns" in the Fed's phony war against Imperial — a war "designed not so much to win the strike, but to avoid any confrontation with the government. With effective picketing banned by the courts, on the grounds it was illegal, this policy was getting the oil workers precisely nowhere." Odam accused the Fed of throwing an affiliate "to the corporate wolves."

On August 16, Ed Lawson of the Teamsters offered to throw his weight behind the oil workers, arguing that the Fed was playing politics. Haines accused Lawson of backhanded support for the Social Credit party, which was always eager for a chance to whip up an anti-union backlash. Premier Bennett jumped at the chance to tie the NDP's leader, Thomas Berger, to lawless unionism. "Wasn't that a terrible thing for Berger to call that bus strike this morning and cause hardship for people without cars?" Bennett asked on August 13. "It's almost criminal to make the workingman who hasn't a car or the pensioner who hasn't a car walk on a wet morning."

On August 25, Berger denounced oil-worker pickets who shut down airport and bus services. He called these picket lines "a continuing act of lawlessness," and attacked the government for not moving in earlier. "The public interest comes first, before the union, and this type of harassment cannot be condoned or allowed," the NDP leader said.

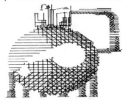

Less than two days later James Harvey was killed. He was 43 years old. A 16-year employee with Shell, he left eight children.

On the evening of August 26, Harvey had put in a shift on the picket line. Nothing much was doing. The union group was taking it easy. Harvey had about six beers. At one in the morning, a professional strikebreaker from Saskatchewan pulled his truck into the yard. Harvey stood in the way. The truck stopped. Then Harvey bent down to tie up his shoelaces. The truck edged forward. The driver's helper yelled out, "Dave, you're going to hit him." The helper testified that he heard a thud as the right wheel of the truck rolled over Harvey, fracturing his head and chest.

The next day, a snap demonstration charged that "Shell Oil supports murder." John Conway, vice-president of the NDP, carried a placard stating, "Death — imperialist method of collective bargaining." Conway called for a general strike.

On August 28, Teamster Lawson pledged that his union would honour all oil-worker picket lines. The Teamsters threatened to go out on strike. "The practice of hot products being brought from behind the picket line by scabs over the dead bodies of oil workers has gone on long enough," Lawson said. On August 29, pickets blocked the U.S. border to commemorate the death. The Building Trades Council called seventy thousand workers off the job at noon. Construction workers paid five dollars each into a trust fund set up for Harvey's family. By week's end, 55 companies applied for injunctions to prevent work refusals.

On September 3, a coroner's jury reported on Harvey's death. It found both the driver and Harvey negligent. There were no arrests, no fines. Both of the politicians who had waxed eloquent against the union lawlessness that had caused so much inconvenience fell silent when corporate lawlessness killed a worker.

The passion around Harvey's death, and the drift towards a general strike that it set off, put the government on the hot seat. Until these events, Socreds had been saying there was no public interest that required government intervention. As soon as the labour minister got wind of the Teamsters' strike threat, he started to rethink. On September 2, he ordered oil companies and the union to submit to compulsory mediation.

Compulsory mediation landed the oil workers in more hot water.

The Fed opposed compulsory mediation. It smacked of compulsory arbitration. It threatened the right to strike. On September 3, 1969, the Fed asked oil workers to boycott the mediation hearings. It offered to boost the picket lines at Shell. Reimer declined to defy the state as well as an industry, unless the Fed was prepared for a general strike. "There's only so many laws you can break at one time," he said.

Compulsory mediation did little to bring company and union closer together. That's to be expected, because compulsory mediation is a contradiction in terms. A mediator's influence flows from the willingness of both parties to compromise. They're willing; they're just shy to tell the other side, in case it's taken as a sign of weakness. So the mediator flies their own ideas past them, and they agree that the ideas make sense. The illusion is that the mediator finds a compromise. The reality is that the parties find a compromise. They just pay a mediator to pretend otherwise.

Compulsory mediation comes from politicians who are taken in by the illusion that mediators work out a solution, when all they do is serve as honest broker for parties that want to settle. Since mediators find workable compromises, all the government has to do is provide a professional compromiser, politicians think. In fact, unless both sides want to get it together, the mediator pushes them further apart. The mediator becomes just another person who needs to be convinced about justice on one side and evil on the other. Psyching themselves up for the uninvited mediator, each side hardens its positions. The mediator is caught in the middle, rather than working the middle.

For three weeks at the B.C. mediation commission, both sides gave no quarter. They didn't have to boycott the commission to make it ineffective. They just had to be themselves. We aren't suffering from a breakdown in collective bargaining, Reimer told the commission. "This has never existed," he said. "The industry thinks collective means a state farm in Poland."

The union demanded a hefty wage increase, $1.25 an hour over two years, higher than the original strike demand. The union said workers needed more because they'd lost so much during the strike. The companies went back to square one. They offered a 50-cent-an-hour increase over two years. They would only deal with a short list of truly national issues. All other matters, including benefits, must be dealt with in regional and plant bargaining, the companies said.

The union made the most of the commission's high profile. Why throw away a chance at free publicity? The commission became a forum for educating members, the public, and the industry about emerging issues in industrial relations. The union tabled two issues — health and safety and pensions — that had never received such attention before. Over 20 years later, both remain at the centre of labour's agenda.

Health hazards in an oil refinery can't be touched or felt, except by scientists. Which is why the issue seemed so "touchy-feely" back in 1969. Unlike safety hazards avoidable by following basic rules, health hazards had no rules except those of the employer. The union wanted an equal say in those rules, and equal access to the research that made rules meaningful. That challenge to management rights touched a raw nerve in an industry priding itself on scientific mastery.

Professor Glen Paulson, a witness for the union, apologized for dealing with such "a delicate matter — fraught with emotional reactions." But, he said, scientists and governments didn't know much about the dangers of carbon monoxide and hydrogen sulphide — both common in oil refineries — except that they were deadly in small doses. He said no one knew how small the doses should be for workers who breathed the gases regularly. The oil companies summoned a refinery manager who said unionists were too emotional about the issue. They shouldn't accompany government inspectors on tours of the plant, he said, because that was a management function. Harry Kavanaugh, a local manager with Imperial, pooh-poohed unions concerns. "When you consider the safety record of a refinery as opposed to the home, you will realize it's safer to be at work than at home," he said. A corporate industrial hygienist from California said industry research was available from the American Petroleum Institute. He encouraged union people to write for it.

That clinched it for commission chair John Parker. Professor Paulson should have known companies were doing this research, Parker scolded. Now that the address of the American Petroleum Institute was known, he asked the union, "isn't it a new ball game? Wouldn't you like to find out from the information given this week what is going on, and reconsider what you want?"

The union also made its case for joint union-company management of pension funds. It was the only way to avoid suspicions that employers were mishandling funds, the union said.

The commission closed on October 6 and recommended wage increases ranging from 41 to 60 cents an hour, a standard settlement in British Columbia and a little better than the pan-Canadian average. The companies immediately agreed. The union held out for 70 cents, which Imperial declared was "just too rich." After five days of give-and-take, both sides agreed to split the difference. At 5:10 a.m. on October 11, they agreed on wage increases from 41 to 70 cents an hour, plus additional shift and overtime bonuses. The particulars weren't that important. The industry had been forced to accept national bargaining -- the underlying, if unstated, issue of the strike —— based on rates set in a high-wage province.

On October 28, Vancouver oil workers accepted the settlement by a vote of 360 to 190. The agreement was extended across the country.

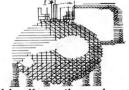

Looking back, Buck Philp says "1969 was the turning point for national bargaining." Within the union, it confirmed that the national leadership could call a strike and settle it. Within the industry, it showed that the union, not Imperial, set wage rates. Tom Phillips, president of the amalgamated Vancouver oil-workers local, felt the difference in relationship right away. "I think the oil companies have a great deal more respect for the men because they know we mean what we say," he told *Union News*.

The settlement whittled away inequality between U.S. and Canadian wages. There was a 64-cent difference in 1949, a 40-cent difference in 1964, a 20-cent difference in 1965, and a 13-cent difference in 1967. After 1969 there wasn't a dime's worth of difference — nine cents to be exact. The settlement also ended inequality in Canadian wages. Newfoundland refinery wages almost doubled.

Reimer still can't understand why B.C. workers were prepared to fight so long and hard for that principle. They could easily have settled for a high rate that left out the rest of the country. It was a noble gesture, he says. "We're just smart," counters Rick White, who was a youthful Vancouver striker in 1969 and by the 1980s was on the union's executive

Women, Children Help Block Deliveries From Oil Refinery

board. In later years, when the economy took a nose-dive in the West, B.C workers got gains set by wage rates elsewhere.

In September 1969, Ray Haines of the B.C. Federation of Labour still wanted to get even. He didn't like the Teamsters showing him up as a pussycat when it came to direct industrial action. He didn't like it when an affiliate turned up its nose at directives from the Fed. He didn't like Reimer's reckless provocation of economic crisis or his appearance before the mediation commission. The strike undermined Haines's reputation and his vision of a strong central labour leadership.

On September 8, at the annual convention, the Fed executive launched proceedings to suspend OCAW. Delegates must decide whether they want a strong central labour body or a "loose-knit debating society," the executive said, charging that recent events "gave comfort to labour's enemies." The report to the convention stated: "Until the Oil Workers Union [sic] allowed themselves to be maneuvered before the Commission through the manipulation of the Teamsters' Union, labour's fight against Bill 133 [compulsory mediation] was extremely effective."

Before a closed session of delegates Haines ripped Philp apart, listing all the Fed policies he'd broken, documenting them beyond dispute. Reimer, not Philp, replied, beginning by pleading guilty to all charges. "Our crime is taking on the biggest corporations in the world. Now if you want to expel us for that, take us on," he said. "We had to find ways and means by which to fight an industry that is as highly automated as they are," he said.

This action was the oil-workers' strike for recognition, on a par with strikes by loggers, steelworkers, and autoworkers from 1945 to 1947, Reimer said. "Over and over again they gave us company unions and company councils, gave their offers and then shoved it down our throats. This time we resolved that we would wreck the council system; and I'm pleased to report to you that we have done that. But the strike had to take its course in order for us to accomplish that."

Reimer continued: "Any act that we performed was in the interest of trying to win a strike, and if we have a crime, it's a crime of fighting companies that are the world over — that operate in as many nations as people we have on strike. You try to have an august body like this Federation on your back at the same time."

The room fell silent. A delegate from the Woodworkers, the largest and most powerful union in the province, moved that the executive resolution be tabled. The motion was adopted, almost unanimously. That shelved the whole matter.

Like so many labour debates over fundamentals, this one was put off to another day. The central issues — the power of a central labour body to lead, and the knowledge and priorities that are required if it is to lead — remained unresolved.

WORKERS OF THE WORLD, RELAX

Canada has a colour bar that segregates blue-collar and white-collar workers. White-collar workers have always enjoyed perks that help them feel superior to blue-collar workers, let them think they belong with management, not unions. They always got to wear nice clothes to clean, air-conditioned offices. They worked straight days on "bankers' hours" or "gentlemen's hours," starting at nine in the morning, not seven. They didn't punch time-clocks. They had an honour system to make up time for midday appointments. In the Depression many of them enjoyed paid holidays and the 40-hour work week, decades before manual workers bargained for those gains. They maintained their lead. In 1973, 82 per cent of office staff worked less than 37.5 hours a week. At that time, despite the fact that most productivity gains were in factories, only 20 per cent of blue-collar workers worked less than 40 hours.

For early industrial unions, the shorter work week became a religion. Overtime pay was meant to prevent overtime, not provide a bonus for it. In 1956, at its first convention, the Canadian Labour Congress declared the 30-hour week a priority. But, despite two-fold increases in productivity, despite the dramatic increase in two-income families, little headway was made on this point over the next three decades. By 1983 less than 20 per cent of blue-collar workers had a 37.5 hour week. In the end unions have accepted the work ethic that is so deeply ingrained in a country where people supposedly learn to work their way up, where a "shop 'till you drop" economy exaggerates the need for possessions, where the "rat-race" dictates "I need this yesterday" and not "tomorrow is good enough for me."

Yet overwork is Canada's top public-health problem. The 40-hour work week — plus overtime, moonlighting, one-hour commutes, and the frantic rush to get to the daycare centre, put the fast food on the table, and get the kids to bed — frays the nerves, drains all energy except for watching TV, makes people sitting ducks for chronic "lifestyle" illnesses, especially heart attacks. The work week leaves little quality time left over for families, friends, or community, which in turn creates a host of social problems.

▼ PART I
▼ SECTION 3 / A STRIKING DIFFERENCE
▼ CHAPTER 5

Shiftwork — standard for most industrial workers — only adds to the difficulties. Folk wisdom has long recognized the dangers of the "graveyard shift." Now, medical research — summarized in the 1986 *American Industrial Hygiene Association Journal* -- confirms that shiftwork not only creates permanent jet lag for natural body rhythms but also causes chronic indigestion, fatigue, and depression. Shiftwork also knocks the living daylights out of relations with friends and family and contributes to drug and alcohol abuse and family breakdown. Yet shiftwork is too hot for health and safety activists to handle. Unions attach a shift premium to it, then insist that health and safety are not for sale.

THANK GOD IT'S THURSDAY

When Neil Reimer and Ted Gaunt of B.A. met on a freezing cold Winnipeg day in December 1969, they knew this time they had to try to get along. With the strikes in 1965 and 1969, oil workers in British Columbia had been on the picket line a total of almost one year out of the last four. Both sides had proved that they could endure long strikes. In 1969 the head of the union and the industrial relations head of the pattern-setting company in the oil industry wanted to see if they could find a better way to reach agreement.

Almost coincident with a change of name — from B.A. to Gulf — the oil company had undergone a change of heart as a result of the union's successful strikes in 1965 and 1969. Gaunt realized, "We should quit fighting for stupid things when you knew damn well the world was changing and we'd better change with it." His associate, Norm Devos, says managers had seen the union as a "passing fancy" until the second strike established that they were here to stay. Now, according to Devos, the idea was that the companies had better find ways and means of dialoguing with the union.

Gaunt admitted he hadn't heard Reimer right when the union director had announced the strike so calmly in 1969. The companies didn't really believe that Reimer could deliver, Gaunt says, that workers would actually risk their mortgages just for the principle of national bargaining. "I wanted to convey that a strike was not emotional, but a calculated and cool decision," Reimer told Gaunt at the December 1969 meeting.

"Would it have helped if I jumped up and down on the table?" Reimer asked. "Yes, I guess it would have," said Gaunt. They agreed to make sure they would try to understand each other better in the future, and that they wouldn't get taken in by the smoke and mirrors of bargaining rituals.

Both sides agreed to start tracking new bargaining trends before they actually hit the bargaining table. For Reimer that involved a new understanding that companies needed lead time to do their homework, think through, cost, and prepare for new programs if bargaining was going to be productive. If bargaining was going to focus on real differences, not just fear of the unknown, unions had to provide company bargainers with some advance notice of major changes. Before managers could take advantage of the advance notice, they had to recognize that union leaders had their hands on the pulse of the workforce, that they spoke for their members, and that their word was good, whether they promised a strike or a settlement.

This new approach to union-management relations was tested in the next round of bargaining. In 1972 and 1973, the Oil, Chemical and Atomic Workers broke the 40-hour barrier in the standard work week that had been in force since the early 1950s —a barrier that no other blue-collar union was able to break during the next two and a half decades. The subsequent failures can partly be explained by the fact that union demands for shorter hours fly in the face of the workaholic and consumer mentality that all North Americans, including (or especially) workers, suffer from; and partly by the fact that other unions didn't have a national bargaining structure capable of pioneering new trends in labour and social relations.

Among union bargainers, there's a saying that "the brass ring only comes around once." There was a time to grab for shorter hours, and that was in the late 1960s and early 1970s. After that the shorter work week became a "frill" compared to the problems of inflation in the mid-1970s and concessions and international competition in the 1980s. National bargaining allowed OCAW to jump at the chance when it was there. National bargaining also produced a leadership style that allowed the union to develop a consensus around a controversial issue: The flip side of the shorter work week, in the short-term at least, is a smaller paycheque. This other issue flew in the face of this society's major motivator — greed.

Reimer sees the 1973 negotiations for shorter hours as "the single greatest bargaining achievement" of his career.

Like most union campaigns, the drive for shorter hours was equal parts happenstance and vision, fluke and design. Like most leaders, Reimer was able to work with all four.

By the early 1970s companies knew their old-time work schedules had run out of time. Absenteeism and turnover were creating havoc as workers scrambled for more free time. That's when management guru Riva Poor -- billed by the Toronto Personnel Association as "the Dr. Spock of the business world" and "high priestess of the four day movement" — turned managers on to the ten-hour, four-day work week. Imperial Oil picked up the idea right away, adapting it for shiftworkers by launching the 12-hour day.

The union was left with an old-fashioned position stressing a shorter workday, while management pressed ahead with the shorter workweek. For commuters and young parents, days not hours counted and it didn't much matter that the number of hours remained the same. "It was worth your life to try to stop those shifts," Reimer says. In 1971 he had written to International headquarters that the shorter work week "is quite enticing to many of the younger people in the plants and where they are a majority we have experienced great difficulty in getting them to reject it." He added, "This is a serious problem and forces our union to develop a posture and program on the whole question of hours of work." Workers were "caught in a vice wanting more money and more time" and "don't particularly care how it is done" as long as the usable free time goes up.

Reimer had to buy some time before mounting an offensive. He warned employers that they were courting a strike if they pushed the question. In 1972, he polled 1,799 members in 94 bargaining units to get a handle on workers' sense of the problems and solutions. The survey gave him a reading on some serious splits. Younger workers wanted longer days and a shorter week. Older workers didn't think they could handle the longer shifts. Although 72 per cent of the members wanted shorter hours, only 42 per cent said they'd strike over the issue.

Reimer decided to move cautiously on the issue. He encouraged staff to sponsor open membership debates. In one mock debate, Ron Duncan made the case for the longer workday, arguing that it would give people a whole day off and force companies to hire more people to fill empty shifts. Reg Basken made the case for the shorter work week based on shorter hours. He said health and safety measures for chemicals are based on eight-hour exposures. Older workers get exhausted after eight hours, and all workers get tired and accident-prone. He said, "Let's not give us medicine on the one hand and pneumonia on the other."

In 1972 the union's annual conference featured the controversy and agreed to leave the matter open. "Members are requested not to stampede their bargaining units into making any decision and cause a split in the union," a follow-up newsletter said in July. "It is important that we have unity in our ranks while approaching the next bargaining program."

In making his approach to employers, Reimer knew he couldn't produce a strike to back demands for shorter hours. "You can't get workers to strike for less money," he says. He looked for a way to encourage the change through bargaining, to persuade managers this was the way to go.

He had already tried his soft pitch on the heaviest hitters in the industry, at the 1973 meeting of the International Labour Organization, a United Nations body sponsored to promote world co-operation on labour issues. The U.S. members of OCAW came out to the meeting to attack Shell for forcing a prolonged strike over workers' rights to sit on health and safety committees. The oil companies came out in full force to contain the damage. OCAW president Al Grospiron acted "like a bear with a sore paw," Reimer says. In a classic replay of the hard cop/soft cop routine, Reimer posed as the voice of sweet reason.

As chair of the ILO health and safety subcommittee, Reimer explained that oil workers had a morale problem because there was nothing distinctive about their product. A carpenter could take pride in a house, but a refinery worker just couldn't take pride in a gallon of gas. So refinery workers needed something distinctive to make them feel good about themselves. Shorter hours were just what the doctor ordered. Shell, fearful of being painted as reactionary on this issue too, went along. So did the other companies. The wheels of the international industry started turning.

Back on the home front, national bargaining put Reimer in contact with the company executives who were in charge of forecasting trends and keeping ahead of them. They weren't stuck in the penny-ante world of plant superintendents. They thought big and over the long term. Before formal bargaining started, probably sometime in 1971-72, Reimer met informally with Ted Gaunt, industrial relations chief of Gulf, appealing to the company's image as a trendsetter. Gaunt liked that idea. He thought the idea of leading a breakthrough was "a big plus."

Gaunt got authority from "the top brass in the company" to go ahead with the plan for shorter hours. He says, "I was pleased because what we negotiated was very important. We broke a pattern. This was a first in North America." Gaunt's team had already "read the markings on the wall," his former assistant Norm Devos says. Devos told Gaunt, "This is not a passing fancy." He suggested to his boss that if shorter hours didn't come in that round of bargaining they were going to become an issue soon after.

"I put it this way," Devos says. "The OCAW at this point in time has to find something that's different than what it's just been doing in the past. The wages are reasonably good, the benefits are reasonably good. What else is there?"

Gaunt also understood that if the company gave shorter hours in the office, they could do likewise in the refinery. Devos used the lead time from early discussions with Reimer to get on top of the technicalities. For managers, eight-hour shifts add up nicely, divide the day into three perfectly-even shifts. The twelve-hour shifts have the same appeal. Using computers, Devos's team cooked up "109 different ways, I guess, as to how you could really shift this." He says, "This was one of the reasons why in our company we were not as opposed as maybe some other companies, because we had done all this work."

Once Reimer sensed management was open to shorter hours (undoubtedly, Gaunt and Devos weren't as forthcoming then as now), he pushed the demand at the oil workers' bargaining meeting in February 1973. Don't vote for it if you don't want it, he stressed, in a negotiator's way of letting the troops know that a victory on this issue would probably exclude other gains. In serious bargaining, a grab-bag of priorities is the same as no priorities, and Reimer needed a firm mandate to focus on the main issue. Membership support gave management the signal that made victory a virtual certainty.

In July Gulf agreed to the 37.3-hour week and wage increases of 22 per cent over a three-year contract. The wage increase, most of which came due immediately, offset the loss of hours and provided a small raise. A new work schedule bundled the shorter hours into 17 extra days off per year, timed to coincide with other holidays. Oil workers call them "Golden Fridays." It was a "historic breakthrough," Reimer told *Union News*, "the first time a company and a union have negotiated a shorter work week with the retention of the eight hour day on a national and industry basis."

If everything looks good, it's because something's been overlooked. Reimer discovered this variation on Murphys' Law the night he negotiated the shorter work week.

Reimer, together with national negotiator Buck Philp and local president Ken Gelok, first convinced Gulf to adopt the 37.3-hour week at the Clarkson refinery. They knew that was just the first step to a national agreement in the Gulf chain, and in an oil industry that followed Gulf's lead. We shattered the 40-hour barrier, Gelok told the local Oakville paper, and it will send a shock wave through industry. Shorter hours will increase employment by 7 per cent, he said, thanks to the union's focus on a social issue more important than "harping solely on wages and prices."

Feeling on top of the world, Reimer and Philp treated themselves to a first-class, all-drinks-included flight back to Edmonton. The phone started ringing off the hook the moment Reimer got home. We've been sold down the river, the callers said. "I went from first class on the plane to steerage in my own house in a helluva hurry," Reimer says.

Reimer hadn't done his homework on the actual hours of work in some refineries. In theory, unions negotiated a 40-hour week and charged premium rates for overtime, and employers had an incentive to hire extra people and thus avoid the overtime penalty. In reality, overtime was part of the normal work week, counted on by both workers and managers. For workers overtime meant more money at pay-time. For employers it meant savings on hiring, training, and supervising extra people. Benefits accounted for about one-third of wage costs. If employers can avoid hiring extra people by paying overtime, they gain a considerable saving.

The 37.3-hour work week threw a monkey-wrench into these arrangements. Reimer had taken Gulf to the limit with 13 per cent increases that made up for 2.7 hours of lost wages. He forgot about other companies where four to six hours at overtime rates also had to be made up. Reimer realized "that many of our members put shorter hours on the table thinking they would never get it." Shell and Texaco, resisting Gulf's lead, offered a 13 per cent increase as long as workers stayed with their old hours. It was a great way to appeal to shortsighted greed and split the membership from the leadership of the union. But the plan backfired. The national agreement was backed by all locals. Texaco faced two strikes — one in Dartmouth and one in Toronto — before giving in to union demands.

No bargaining breakthrough is an island. If a new gain isn't extended, it stands out like a sore thumb: a competitive disadvantage and political embarrassment to the company that gave it up unnecessarily. It won't be long before it's withdrawn.

"I think that point sometimes gets lost in bargaining, that we build up IOU's to companies" that agree to pioneer, says Stu Sullivan. In 1973 and 1974 Sullivan was the OCAW staff rep in charge of extending the shorter hours victory to "chemical valley" in Sarnia.

By pure chance, Fiberglass was the first Sarnia plant due for bargaining after the oil pattern was set. It wasn't a strong company to lead with. The plant had a 42-hour work week, which would make the sudden drop to a 37.3-hour week hard to accomplish in one step.

Lacking strong support from the local bargaining committee, Sullivan relied on sheer bluff and the threat of militant strike action. At one point in the bargaining a local leader showed interest in the company proposal of a 40-hour week. "I never kicked anyone so hard under a table," Sullivan says. "That shut him up and I moved right in and explained that what he meant was that we could have looked at the proposal two years ago, but now it was ridiculous." The company agreed to the union demand in an eleventh-hour settlement mediated by the ministry of labour.

Sullivan repeated his success at Polysar, and at Cabot in Welland.

Then giant Dow Chemical took the union on in October 1973, and 819 workers went on strike. Bill Maughan, a seasoned local leader with 28 years' experience in the Dow plant, thinks managers "hung their hat" on the hours issue, but that was really a side issue. "I think they felt this was the time that they were going to break the union," he says. They brought in supervisors and material and got ready for a long siege. It was the same game-plan they followed successfully to rid themselves of unions in the United States.

As Sullivan saw it, the Dow strike couldn't be won on the hours issue alone. Although workers wanted shorter hours, "It wasn't a major enough issue that we could let it stand by itself, so we frantically had to develop some other issues." The union's bargaining committee worked up 13 thorny issues. The most popular called for reinstatement of Telfor Adam, a pipefitter demoted for causing a chlorine spill. The plant had wildcatted when Adams was demoted. Workers felt Adams, a Native, was being discriminated against and scapegoated for the spill, which was an accident that company managers frequently caused as well. Gordon Burthwick, a member of the local executive, says "That's what banded us together because people said:`what the devil are we on strike for a shorter workweek for? That's less money.' But it was that one guy that really banded us together and made that damn strike work."

Three weeks into the strike Dow offered to shorten the work week if a plant referendum showed workers truly wanted it. If the majority favoured 37.3 hours the company would agree, managers said. This was a clever manoeuvre. On the surface it offered shorter hours, a short strike, and direct democracy. For the union it was a death threat, a ploy to bypass the union as sole bargaining representative of the workers. If the company got the go-ahead to appeal to workers over the head of the union's elected bargaining team, it would be one step closer to a non-union shop. Then the company offer could be withdrawn without penalty, because no contract would have been signed.

A raucous membership meeting debated the offer, but workers backed their union. From then on, the strike was "down hill sledding," according to Sullivan. Times were good, and pipefitters and welders had no trouble finding part-time work. They donated part of their earnings to the strike fund. The local made the strike a family affair. Sullivan turned a blind eye to the local's informal "sex committee," a swingers' club with a mandate to date the wives of the strikebreaking foremen who remained in the strikebound plant. Once rumours of the committee's successes got back to the plant, "The next thing we know, there's a hell of a lot of pressure put on by the supervisors to end the goddamn strike so they could get home," Sullivan says.

Though Dow's unionized workforce was all male, Sullivan called a meeting for wives to discuss their concerns about the strike. He invited them to help in any way they wanted. "The only things I excluded them from was cleaning up and cooking at the strike headquarters," he says. The meeting formed a women's committee that took charge of a daily strike bulletin to spread the word to the community. The committee studied the legal injunction that outlawed mass picketing at the plant gates. It said nothing about a parade, so the wives' committee got city approval to organize a mass parade on the plant gates.

In December Dow gave in to 12 of the union's 13 demands. The wives weren't keen to settle. They wanted their husbands home for Christmas. In response to their pressure, the union agreed to approve the company offer only if individuals remained free to make their own choice about working the holidays. Over eight hundred people had been on strike for 13 weeks, and Christmas work paid double time. Not one worker volunteered to return early. Sullivan says, "They told their supervisors, `You wanted to run this plant; run it over Christmas for me.' It was the wives that did that."

Shortly after the strike, in March 1974, OCAW locals set up new union offices in Sarnia. The labour temple provided space for the local labour council as well as OCAW functions. The basement was turned over to a food co-op, another floor to a lounge and dance floor. The building remains a monument to the family and community ties strengthened by the struggle for the shorter workweek.

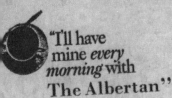

NEIL REIMER OF THE OIL, CHEMICAL AND ATOMIC WORKERS INTERNAT
... as much a part of the oil industry as the management he has batt

Neil Reimer:
The man the o
barons fear m

Neil Reimer belongs to that small elite whose name becomes synonymous with a certain event, cause, action or group. Reimer's field is in

church and the state, couldn't

By Paul Jackson
of The Albertan

Credit government's attitude
Labor as "insidious."

OIL ON TROUBLED WATERS

Until 1969, national bargaining was an OCAW strategy to gain union recognition, to stop companies from playing one region off against another, to put the union in the driver's seat of bargaining trends. According to Neil Reimer, when he and Ted Gaunt met in 1969 and recognized each other as the industry's pacesetters in bargaining, "We both had a sense that if we couldn't beat the other guy in the alleys, it made sense to start working together."

Later, national bargaining was put to different, more complex tests. The mid-1970s featured runaway inflation as raw material prices went through the roof. The early 1980s featured runaway recession, as oil and chemical prices tumbled and brought down a host of over-extended companies with them. National bargaining had to weather these economic ups and downs, when turbulent change was pressuring most unions on the continent to cut a deal and run. In the process, national bargaining proved itself as a vehicle suited to both rank and file activism and creative tension with management.

127

Who's Ever

EDMONTON (Staff
product may have a
it, but there's nothi
ing when it comes to 1
picture.

So say a group of
ton businessmen who
rently launching a na
doughnut franchise
tion.

While the firm a
has only one store
tion, Dolly Donut Ca
is now processing
30 applications for
outlets in places as
as Saskatoon, Sask
Francisco, Calif.

Says Ed Symba
pany president: '
we have a winner
face it, what ta:
than a fresh doug
cup of coffee?''

Just in case you
keen on the aver
nut, Dolly Donut
thing else to hool
Most outlets will
thing like 40 dif
of the product.
really like a
choice, Symbalu
firm can even c
different types,

U.S. reserves of
are depleting

seems strange
he U.S., when it
nething such as
atal oil policy,

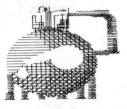

If high wages caused inflation, unions were the last to find out. The cost of living went up 3.1 per cent in 1971, 2.9 per cent in 1972 and another 4.8 per cent in 1973. Most unions conducted business as usual. They signed two-year and three-year contracts, confident that the dollar would stretch out that long. When inflation took off, jumping to 7.6 per cent in 1974 and breaking into double digits — 10.9 per cent — in 1975, unions coming out of long-term agreements had a lot of catching up to do. Ironically, their catch-up settlements left them open to blame for causing inflation.

Oil workers got caught at the tail end of this spiral. In 1973 they focused on the shorter work week. Most of their wage increase just made up for the loss of hours. To offset the paltry increase in take-home pay, Reimer had the agreement "front-end loaded." That is, most of the raise came in the first year. Hardly any increase came in the second year, when prices took off.

Nor were their wages protected by a Cost Of Living Allowance (COLA), often called "the clause that refreshes" because it automatically adjusts wages to inflation. The Autoworkers are long-time champions of COLA. Reimer is a long-time opponent. In the 1950s he terminated any COLA clauses left over from wartime. He thought it was morally wrong. "You can't build pressure to stop inflation if the best-organized workers are rooting for it," he says. And COLA formulas, all pegged to percentage increases, inevitably discriminate against poorly paid workers, for the simple reason that 10 per cent of ten dollars an hour is more than 10 per cent of five. Reimer also thought COLA was a weak bargaining tactic. He wanted to base his argument for wage increases on productivity improvements, not the rising cost of living. He also recognized that COLA clauses came at a high price. Under COLA agreements, companies lose their ability to predict labour costs. To protect themselves from the unexpected, the companies hold back more in bargaining.

Reimer's approach to COLA worked as long as the cost of living remained fairly stable and predictable. He took an extra precaution in 1973, in the event that prices didn't hold steady. He had an informal understanding with Ted Gaunt that Gaunt would use his influence to reopen the contract for wage adjustments if inflation took off. That "deal" became public early in 1975 when an OCAW organizer promised non-unionized Gulf workers in Edmonton that a wage-reopener was a sure thing. Join a union that can reopen its contracts, one leaflet bragged. Local managers immediately denied that any such deal was in place, and Gaunt couldn't contradict them. The deal was off.

It was about this time that Reimer refined his personal theory that a collective agreement was a "living thing." He waxed eloquent about contracts that grew with the times. The point, he wrote company heads in late 1974 and early 1975, was that they should apply their green thumbs to a growing document by topping up current wages by $1.50 an hour. Companies could "avoid all kinds of difficulties by making a realistic wage offer," he suggested. But the oil companies didn't volunteer to do this.

So Reimer issued a bargaining bulletin to workers outlining his theory of a living agreement. Companies that make record profits but won't "deal realistically with an interim wage increase," he wrote on January 3, 1975, "will undoubtedly have dissatisfied employees and experience low morale and difficulties with them." On January 6 he hastened to add, "It is a continuing OCAW policy to honor our collective agreements." He called a special bargaining conference of oil workers, to be held at the Holiday Inn in Edmonton in March. At the meeting he explained that although it was his duty to inform them that wildcat strikes were illegal, he sympathized with their frustration, and understood why so many of them were considering protest actions. Wink, wink, nudge, nudge.

What Reimer said about collective agreements being living documents was true. Though the agreements are legal and binding, the word "contract" is a misnomer borrowed from the business world. Contracts cover purchase agreements and parties agree to live up to the letter of a contract as soon as it's signed, sealed, and delivered. Collective agreements are constitutions that deal with human relationships. They have to deal with the unwritten law, the give and take that grows out of a sense of fairness and exchange. That's why coffee breaks always take longer than the rule book says. That's why workers let certain formalities slip so they can get the job done, why "work to rule" is only followed when workers want to throw the book at an officious employer. Wise managers and union leaders know not to press a technical advantage too hard, lest they ever find themselves at the receiving end — in a jam and needing a little technical assistance. Such moments usually come when they're least expected.

On January 17, 1975, Texaco in Montreal got set for a "turnaround." That's when a plant shuts down for major repairs and everyone has to pitch in, work flat out, because everything's down, there's no production. Every minute counts, every person counts. It was a perfect time for a wildcat strike. No sooner did the Texaco workers walk out than BP refinery workers joined them in solidarity. Hank Gauthier, the staff rep, got BP workers to picket Texaco — and Texaco workers to picket BP. That way, no one picketed his own plant during the illegal strike. Live by the book, die by the book.

Refinery workers elsewhere in the country found themselves unable to come to work for a day. But in Montreal the workers refused to go back. They stayed out six weeks. BP and Texaco refused to bargain an increase because that implied acceptance of an illegal strike. Gaunt, who was in touch with Reimer and Philp, advised the companies that a 12.5 per cent increase would settle the matter. That still didn't solve the problem of saving face. Texaco refused to sign an agreement with a gun to its head. The strikers refused to face the jeers of supervisors if they didn't go back with a win. Finally, Gauthier got Texaco to send its supervisors home for the night and the workers returned to work pretending there wasn't an agreement. Then the company voluntarily offered a wage increase. Gauthier says, "I revelled at that. It was a nice piece of engineering."

Then Texaco and BP went after the union. Each sued OCAW for illegal strikes that had cost a million dollars in lost production. Reimer asked his lawyers to buy a little time. "They can take an awful lot of time in discoveries and what-not, those lawyers, especially if you're paying them," he says. As the companies' spite wore off, his lawyers negotiated a reduced penalty of one hundred thousand dollars for each company. The companies also wanted a letter of apology. "That'll cost you," Reimer said. Each company got a letter and $75,000.

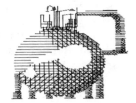

In 1974 Prime Minister Trudeau was re-elected on a promise to "wrestle inflation to the ground" without resorting to the Conservatives' program of wage and price controls. He called wage and price controls "a proven disaster looking for a new place to happen." After the election inflation turned to "stagflation," double-digit inflation without corresponding economic growth. Inflation ran at 11 per cent. The unemployment rate jumped to 7 per cent and interest rates went to 9.5 per cent. And the second-highest strike record in the world netted Canadian workers, most of whom had been without new agreements for two years, average increases of 18 per cent. In October 1975 Trudeau slashed the federal budget by $1.5 billion and imposed wage and price controls. To further "cool the fires of inflation" he set up an Anti-Inflation Board (AIB), which would reject bargaining gains of over 10.2 per cent or $2,400.

OCAW, like most unions, called Trudeau's program "wage and wage controls." The AIB budget encouraged a major hike in oil prices, to jack them up to world levels. The prices of imports weren't controlled. The wage ceiling was below the rate of inflation.

An OCAW bargaining conference in January 1976 voted to bargain one-year contracts to minimize the damage of the AIB. Delegates also voted to turn the AIB's ceiling into a floor. All locals were told to strive for the maximum allowed under the law. Strikes were discouraged. They wouldn't be productive as long as the government could roll back "excessive" increases, delegates decided.

Then the mood became more militant. In February 1976 Reimer issued strict instructions that no local was to accept any wage increases below 10.2 per cent. He also offered full support to any local that defied AIB rulings. In March he announced that the union would appeal AIB decisions and would fashion its agreements using loopholes in the laws and regulations.

Workers at Husky Oil in Lloydminster, Saskatchewan, struck for national rates from November 1975 to February 1976 and achieved them — after a decade's failure to do so with a company that was owned by Mormons and was anti-labour. Because of the union's defence of national bargaining at the AIB, the AIB let the agreement through.

No government had interfered with free collective bargaining on the scale of Trudeau's AIB since World War II. Which is where Reimer had the edge. He was one of the few union leaders who had gone through wartime controls. He remembered how unions and companies had colluded to beat those controls. He says, "You had to go to the board with the employer. You had to argue that you'd face a labour shortage if wages weren't increased. You had to argue that there were exceptional new job duties that justified a bigger raise. You had to argue that there were historical relations with other groups that had to be maintained." He learned all this from watching the companies operate. "They were the first to catch on to these arguments when they went for increases for their supervisors."

Quite independently of wage controls, the union had been moving to its own version of classification upgrading. In 1971 Sarnia union rep Stu Sullivan took a fistful of grievances to Vince Norwood, the Shell refinery manager. Everyone wanted their holidays rescheduled. "Do you think you can do any better?" Norwood asked. Wrong question. Instead of grievances, Norwood got a union-management task force that said workers should figure out their own holidays. When that worked out fine, Sullivan proposed the same way of handling disputes over training opportunities. When that worked out fine, there was a training bulge: too many skilled workers for too few skilled jobs. Simple, Sullivan said. Pay everyone according to their knowledge, and assign them according to seniority and the needs of the moment. Junior workers who'd taken training for senior jobs got the senior rate of pay, regardless of their assignment.

The AIB was the perfect time to apply this kind of lateral thinking to national bargaining. Traditional managers try to save on labour costs by paying different rates for jobs at different skill levels. Traditional unions try to make that as fair as possible by giving senior people first crack at the best jobs. The system is "penny-wise, pound-foolish." Managers lose out on the flexibility that comes when workers can handle a variety of jobs. They have to pay extra for overtime and call-ins if someone is sick, which is a major headache for plants that function on 10-hour and 12-hour shifts.

"And from the safety point of view, you'd think the owners of a $300 million plant would like the comfort of knowing they have some depth of knowledge in their plants," Reimer says. The traditional system is also hard on unions, which become the bearer of bad news to junior workers who have to wait their turn for better jobs and pay-rates.

About one-third of union grievances deal with workers complaining that other workers got the job they wanted.

OCAW negotiated another system based on "pay for knowledge." Managers would get a range of skilled workers to draw on, and workers a chance to better themselves and try out a variety of jobs without having to wait for the high-seniority people to retire. It's called "win-win" in the trade.

The system encouraged flexibility and training. It raised average wages in a plant by $1.50 an hour. And the union won the right to a joint say in training programs. There is no record of a seniority grievance in any plant that has adopted the system. Indeed, the system allows seniority to work, because it gives all workers the chance to learn skills at their own pace.

That method of achieving pay raises allowed the union to break the AIB guidelines. Reimer, following the wartime strategy he observed in his youth, made a joint union-Shell presentation to the AIB, knowing that Shell enjoyed more clout than the union. Shell had donated some of its executives to the AIB to help it in its deliberations. Superior wage increases were justified, the union and company argued, because they were based on new duties outlined in the new pay for knowledge plan. "We did well with this type of argument for the first two years of the AIB, but not the third," says Reimer. "We couldn't bullshit three years in a row."

Co-operation with companies was limited to joint appeals for wage increases that went beyond AIB rules. Otherwise the union maintained a traditional anti-corporate stance. In February 1977 OCAW asked the AIB to roll back gas prices and "establish a rational energy policy based on the needs of Canadians, not of the multinational corporations and a few greedy politicians." In fall 1977 OCAW locals joined the "national day of protest," the largest general strike in the country's history, to demonstrate against wage controls. For the most part, companies were given advance notice of the shutdown, and skeleton crews were left behind to make sure equipment wasn't damaged.

These statements and actions give the measure of the union's fundamental conflict with the oil industry. Their differences aren't based on nickels and dimes. The labour costs of refining a litre of gas don't amount to two cents. Companies don't oppose unions to keep workers from getting their two cents worth. The enduring basis for conflict in the industry is political, it has to do with the questions of workplace, social, and economic organization.

Lines of conflict on the shopfloor, for instance, aren't drawn by the production process itself. Refineries aren't assembly lines where supervisors serve as overseers who push workers to produce, where the differences between bargaining unit and management duties are hard and fast. Operators in refineries and petrochemical plants have all the knowledge and control they need to run their own show. They work as a team and co-ordinate and discipline themselves. Slackers have a hard time getting workmates to swap shifts when they need a weekend or night off. Until the 1960s, take-charge personalities from the shop-floor commonly worked their way to the top. John Stoik, later chief executive officer at Gulf, got his start as a worker and then superintendent at the B.A. plant in Moose Jaw. Polysar headhunters recruited two top executives, Tom Towler and Frank Hubbard, from the union's leadership. Bodysnatchers at Latex turned former chief steward Bill Findlater into a senior manager. On their way up, they had to take their lumps from cheeky operators. When Stoik became plant supervisor in Moose Jaw he had to deal with Stu Sullivan's pestering for air-conditioning in the control room. Stoik didn't see the need for such luxury until he made a routine inspection in mid-July. Sullivan and his buddies kept the supervisor busy with their questions. They watched intently until Stroik inevitably loosened his tie and wiped the sweat off his forehead. Then came the last question: Have you reconsidered our request for air-conditioning? Sullivan says, "He just looked at us and said: `You bastards!' Three days later, we got it."

Later, oil refineries started to recruit executives straight out of business school and got sticky about management privileges. Sullivan says, "They put a man in charge of 300 million dollars of equipment, and if there's a cat reversal, he has 45 seconds to react or the place will be blown sky high. He doesn't have time to consult with the supervisor." Sullivan, who became a senior union staffer, says the companies trust a worker with that kind of decision. "But the same worker has to get the foreman's signature on a meal ticket for overtime." And now, with advancement closed to shop-floor workers, there was no escape. "That's why oil workers are the biggest whiners in the world," Sullivan says.

It was management, not union organizers, who made a big deal about the status of supervisors. Their role was manufactured by management, not by the needs of the production process. When Roy Jamha tried to organize a new Alberta oil refinery in the 1950s, management classified all its supervisors as bargaining unit members, counting on them to tip the balance in a close vote. The union won the vote anyway. "We made them suffer for that one," Jamha says. "I think we got about ten years without a supervisor in the plant, and it was the best thing we ever did." As the numbers of supervisors swelled during the 1960s, they were viewed as a management power play. According to Sullivan, "They're just a standby force of strikebreakers," a practice that is common in highly automated monopolies. The supervisors hold on until major backlogs or emergencies grind production to a halt. By that time, if management's theory works, workers are desperate to settle.

Reimer's sense of labour relations grew out of this shop-floor recognition of management's social and political, as distinct from crude cost-cutting, motives. Throughout the 1950s he heaped scorn on the "Santa Claus complex." Paternalism is wrong, he told the Alberta Personnel Association in May 1957. "The employee should be allowed to help himself." That was a swipe at Imperial. Reimer found it almost impossible to organize when Imperial non-union plants matched the wages and conditions of unionized companies. The experience gave him a lifelong contempt for paternalistic policies that reduced workers to recipients, not equals.

Labour relations were warped by this every-inch-of-the-way battle just to establish a union's right to exist. When K.C. Irving tried to crush the union in 1963, workers rallied to save their union. "In labour relations, it's funny. It's an interesting science," Reimer says. "People haven't figured out how to get along. But everybody knows how to fight. So, if the employer tries to crush the union, there'll be an equal response from those who fight back to save the union." The harsh struggle for sheer survival spared no quarter for new ways of thinking. "It was just go ahead, and do what we did yesterday. And the only way out is real recognition."

In 1963 William Dodge of the Canadian Labour Congress suggested that union leaders co-operate with managers to work for full employment and a rising standard of living. Reimer delivered a blistering attack on the scheme in a press release. Managers couldn't be trusted until they recognized unions, seniority, and other basic principles of industrial democracy, he said. "When management demonstrates at the work-a-day plant level that it fully recognizes the rights of workers to organize and that it is considering at that level the human elements in its economic and mechanical calculations, then I will recommend that we co-operate at the top level. Until this happens, I suggest to all locals that they be on guard against high sounding but vague talk about labour-management partnership." In a separate release to union locals, Reimer stressed that it was "very dangerous to have co-operation on a National level, without the corresponding demonstration that Management wants to co-operate with the workers" at the local level. He warned that unions "have been trapped and taken advantage of" by this in the United States.Once the 1969 strike secured union recognition, however, Reimer tried to use national bargaining to open doors to chief-executive offices. He had a hunch that executives would be more open to reason than their underlings, take more interest in clearing up needless misunderstandings. "They have a greater incentive to make the place work than the guy coming up the ranks who wants to succeed them," Reimer says. "And it keeps the junior managers honest, makes sure they don't colour stories that go to the top." He was groping for a way to break through the limits of collective bargaining and start the process of negotiation.

The words "bargain" and "negotiate" are often used as synonyms. They're both midway on the spectrum of methods for addressing conflict. But bargaining leans in the direction of feud-struggle-argument. Negotiating leans towards exploring needs and priorities-problem-solving-dialogue. They each involve a different style, process, and method of getting to the nub of problems and solutions. They also bring different issues to the fore.

Bargaining and negotiating use equal doses of persuasion, pressure, conflict, and compromise. But bargaining goes with a push-and-shove relationship. Negotiating goes with give-and-take. Bargaining sets a price tag that can be finalized in a contract. Negotiating sets out problems that require ongoing debate. Wages, hours, and seniority can be bargained, but productivity problems and frustrations with dead-end jobs have to be negotiated. Some industrial relations consultants like to package these differences under "win-lose" or "win-win," as if the difference lies in the method, the outcome, or the desire to collaborate. The real difference lies in the relationship, which is much harder to sustain than new styles of conversation at the bargaining table. Bargaining can produce a package. Negotiation has to produce a process. Negotiations require an open-ended relationship and ongoing exchange about values, directions, and priorities.

Canada's industrial relations system tolerates bargaining but discourages negotiations. Labour laws, for instance, are premised on the need for stability, not justice. That's why governments outlaw "wildcat strikes" and force unions and managers to settle differences through the legalities of arbitration rather than through learning to work out the issues or face the consequences. There's no charter of rights that guarantees free speech — or participation in decisions about work.

Local bargaining is biased against negotiations. Wages, fundamental to any agreement, have to be bargained from scratch each time. As deadlines approach there's no time for other issues, which get left behind to fester, to become ideal breeding grounds for continuing feuds.

In 1974 Reimer decided to press his luck on the negotiation front. He wrote corporate chiefs to warn them that they better start finding ways to work with unions to make bargaining work; otherwise the government would intervene and impose its own solution on runaway inflation and strike rates. Later he invited corporate presidents to address sessions of the annual union staff educational retreat. That was too much. Reimer says, "They no longer said, `That's Reimer-reason'; they said I'd gone senile early." The corporate heads turned out to the sessions but they were nervous. Dick Clarke, president of Celanese, told the 1980 session that he'd never met a unionist before. Alex Hamilton, an old fighter pilot who headed Domtar, told the 1981 session that he'd just finished helicopter skiing in Banff. "I'm a hell of a lot more scared today than yesterday," he confessed, "because in my whole career I've never sat down with a union leader."

In 1978 a top Shell manager spoke about the workplace of the future, where managers would have to give way to "less supervisory control and more self-control within the groups." The new environment could make bargaining more rational, but not less adversarial, he said. "The current management of Shell Canada is not of the school which equates union weakness with management strength," he said. "We are generally aware and painfully conscious in some cases that weak unions tend, in place of resources and leadership, to use irresponsible demands, demagoguery of an ideological nature and the revival of old realities and myths which have nothing to do with present-day labour relations."

The 1970s were good years for the oil and chemical industry, and talk of dialogue was cheap. But by the early 1980s the oil and chemical industries were locked in a world recession. Gaunt had retired from Gulf. The new chief bargainer, Norm Devos, didn't have Gaunt's clout. The company had recently reorganized along product lines, and the overarching industrial relations department had lost its free hand. The company opened bargaining with demands for union concessions and a return to the 42-hour week. The company argued that it couldn't keep up wage levels set in a more prosperous era, that non-unionized office workers had all accepted concessions gracefully. The company wasn't prepared to talk until the union also accepted the principle of the need for concessions.

Reimer was pledged to refuse any rollbacks of previous gains. But he didn't want to be manoeuvred into a single-issue strike against an industry that suffered from excess capacity. So he bit his tongue and smiled. "Half your time in bargaining has to be spent stopping your adversary from digging a hole for himself," he says. "When it looks like they're going to say `never,' I switch the topic to the weather."

Reimer knew that one thing that was on all executive minds was the need for efficiency improvements: they all knew that at least one company was going to fall by the wayside in the new era of oil glut and falling prices. He also felt the union was on strong ground in a discussion that centred on efficiency. He had never based his arguments for higher wages in the 1970s on the high price of oil. "I never thought it had an economic basis," he says. The oil price "was a political price, designed to raise money for the war machine in the Middle East. And as soon as they saw new areas coming into production, they lowered their prices to put their competition out of business." So he continued to base his wage arguments on increased productivity, which at least gave him the virtue of consistency once prices collapsed. Reimer also believed that the industry was inefficient. During wage controls, he says, companies inflated their managerial staff so executives could base their case for higher salaries on increased supervisory responsibilities: the Pierre principle — work expanded to meet the control loopholes available.

"So," Reimer says, "I countered with a proposal for a joint dialogue program on internal efficiencies. I said I'd let the chips fall where they may; if the responsibility for inefficiencies lay with us, we'd make adjustments; if it rested with them, they would." Good negotiators, like good lawyers, don't ask questions unless they know the answers.

Just to be sure, Reimer circled Gulf by bargaining no-concession contracts with co-op refineries in the west and with BP. BP had always been a follower in the industry, but as the company's managers prepared to be swallowed up by Petro-Canada, "They decided that since they'd never made a contribution before, this was the time for a last fling."

Reimer wasn't prepared for the crisis his proposal provoked in Gulf executive suites. The company took three weeks to respond. Management "hawks" didn't like the sound of it. Inviting the union to comment on management's efficiency was out of bounds, they said. Bill Baker of Gulf's industrial relations team was open to the idea. "But on the other hand," Baker says, "we weren't the ones with our feet to the fire." The idea "wasn't received as badly as I thought it might be," Devos says. Gulf's new president wanted a breakdown on productivity problems, so the proposal fit in with his agenda. The hawks lost the argument. Many lost their jobs as well.

When the "doves" won out, Reimer responded in kind. He appointed top union staffers Buck Philp and Reg Basken to the four-person task force. "We were flattered," says Baker, who recognized the signal that the union would be serious and constructive. Senior executives knew there was a problem with junior management inefficiency, Philp says. "For years and years, money was nothing, so they never bothered listening to people on better ways of doing things. They just plugged along in their own way, regardless of cost." As the reality of hard times filtered down, "It made them aware they had a problem and they'd better listen."

At the Port Moody refinery, the task force found a heater that wasted $50,000 a day. No one had bothered costing it before. Repairs were easily made. Tool cribs were also reorganized to cut down on lost time spent hunting for tools. The same procedure uncovered $3 million of waste a year in Edmonton. The exercize eliminated the need for concessions. But union input stopped there. Managers weren't ready to give up control. "They talk big about Quality of Working Life," says Philp, "but when it comes down to the nutcracking, then it reverses."

Negotiating, when all was said and done, was only midway on the spectrum towards genuine problem solving. The national bargaining experience confirms that the toughest nut to crack is power sharing, and that the greatest fear of energy companies is their own power failure.

GOD'S FROZEN PEOPLE

Heavy oil in the north has been a gold mine for academics and journalists in the south. They've zeroed in on the loners, misfits, and transients who make up the wilds of Fort McMurray. They've sensationalized the sin city of the Athabasca, far from the civilizing effects of Toronto. According to the *Toronto Star*, citing RCMP officers in a typical story of April 1977, the town is a hideout for criminals and gamblers.

Of course, any town that grew from a frontier outpost of 2,600 in 1961 to an industrial city of 31,000 by 1981 will lack some elements of peace, order, and good government. The people who volunteered to become God's frozen people weren't established citizens. They were the young, huddling to be free. In 1980 half the population of Fort McMurray was under 25. The average resident had been out of work twice and switched jobs five times. Some five thousand citizens were refugees from unemployment in the Maritimes. Without regular family and community support systems, many people in the town, especially housewives, suffered from cabin fever.

They came to make ends meet, not to fall off the wagon at the end of the world. Crime and drinking levels in Fort McMurray aren't out of whack with other places, Charles Hobart reported in the 1978 *Canadian Journal of Criminology*. Surveys by John Gartrell for the Alberta government showed that close to half of the newcomers in both 1969 and 1979 came with intentions of staying for some time. Less than one in five came to take the money and run within a year. Home ownership rates were the same as elsewhere. Most people belonged to at least one club or association. One in four did volunteer work. Gartrell's study, reported in the 1983 *Canadian Journal of Community Mental Health,* found no abnormal signs of stress, alienation, or powerlessness among men. Wives — 40 per cent of them had no choice in the decision to migrate — were worse off.

The town's make-up was harder on union organizers than on social workers and police. Turnover in both the Suncor (originally Great Canadian Oil Sands or GCOS) and Syncrude plants was high. Most workers were young and from

ALL'S QUIET ON THE WESTERN FRONT

small-town backgrounds. They were blown away by the high wages and overtime bonuses, all part of isolation pay. "This was their one chance to make some money and they didn't want to risk it," says Peter Gorrie, co-publisher of the Fort McMurray *Express* from 1979 to 1981. The companies made it clear, he says, that unions would bring strikes and layoffs.

Gorrie supported the union drive and got a lot of anonymous tips about health and safety problems at work. He says, "People were really worried about their lack of control, but their lack of control meant they didn't want to rock the boat." They saw the union as controlled by outsiders. "I wouldn't have predicted that they'd unionize," Gorrie says.

In the fall of 1967, five hundred dignitaries cracked open the champagne to toast Northern Alberta's tar sands, launch site for the world's first oil mine. "A lonely looking old man sat silent and impassively at the head table," author Larry Pratt reports, "huddled deep into a blue overcoat, the collar turned up at the back and rimless spectacles riding down an ample nose."

The mystery guest was John Howard Pew, chief of Sun Oil, the main man behind the Great Canadian Oil Sands venture. A stern Presbyterian, according to company historian Arthur Johnson, Pew bankrolled Bible colleges and the Christian Freedom Federation. A holy roller for free enterprise, he also financed the J. Howard Pew Freedom Trust and a variety of far-right political lobbies. Pew treated his employees like a good father, another company historian, August Giegelhaus, says, and promoted company unions, "a combination of Sun paternalism and active resistance to organized labour," to keep real unions out.

As a right-wing evangelist, Pew was a

139

man after Premier Ernest Manning's own heart. As an investor, Pew's interest in Alberta's tar-baby was the heart and soul of Manning's labour and economic strategy. Manning knew Alberta couldn't match the Middle East's cheap oil, but he trumpeted Alberta as an oasis from the shifting sands of Middle East politics. "Accustomed to dealing with fermenting political climates in insecure areas around the world, the industry has always found a welcome contrast in stable, conservative, prosperous Alberta," Manning told *Oilweek* in 1962. A "union-free environment" was part of the red carpet treatment for U.S. investors. Alberta's labour code of the early 1960s had the down-home feel of the Southern United States. The code legitimized company unions, forbad unionization of professionals or technicians, and gave the labour board discretion to stop strikes and quash unions.

The Pew-Manning axis was bad news for Neil Reimer, director of the oil workers' union. Reimer says, "We had to be established in the oil sands for national bargaining to be credible." He knew the future of the oil industry would be decided there. There's enough oil in the gooey sands, Joseph Fitzgerald writes in his 1978 book *Black Gold With Grit*, to pave a four-lane superhighway to the moon.

Early proposals to get at the oil were also out of this world. In 1959 the federal cabinet discussed beating the tar out of the sands by dropping a nuclear warhead on the area. When the politicians came down to earth, they realized that mining the oil sands was going to be labour-intensive. They relied on Manning-style anti-union laws to keep a tight rein on the unbridled forces of a frontier boom. Employers don't like being "held for ransom" by workers who know that labour is scarce, deadlines are tight, and delays cost a fortune. To lure investors to the tar sands in 1973, Pratt shows in *The Tarsands: Syncrude and the Politics of Oil*, government negotiators promised a no-strike pledge. In Fort McMurray, according to a Gulf newsletter, job-seekers were told that the plant "will be operated without unions." When the union saw this, it charged the company with unfair labour practices — workers, not bosses, are supposed to decide whether a union is wanted — only to have the labour board rule that the employer was just exercizing free speech.

The long open season on unions had its effect. Alberta became the stronghold of "company" or "independent" unions.

Reimer hoped to throw a surprise party in time for the grand opening of the Great Canadian Oil Sands in 1967. In March he tried to organize the skeleton crew preparing the mine site for the start-up. He flogged union cards in the open while an Edmonton chemical worker hired on as "bullcook" worked from the inside. They signed up 80 of the 130 workers, enough to win the vote, until the company hired on 40 extra workers and spoiled their plans.

Stu Sullivan took a second crack at organizing in Fort McMurray in 1968. He quickly figured out that he couldn't win with traditional union appeals. Most of the workers were fresh from subsistence farms in Northern Alberta and were making big bucks for pretty soft jobs. Sullivan came across as a spoil-sport.

Sullivan soon met a welcoming committee in the person of Mike Woodward, leader of the company-sponsored employee association. "I'm going to pound the shit out of you," said Woodward, a former boxing champion in the armed forces. "That's the coward's way out," Sullivan replied. "Why not listen to what I have to say?" That stumped Woodward, temporarily. "But if you don't convince me, I'll pound the shit out of you later," he said. Woodward signed up with the union that night.

Woodward and Sullivan decided on a community-based organizing campaign. Workers who came from small, stable societies may have liked the pay but they had trouble adjusting to the hectic pace and anonymity of a boom town. They wanted neighbourhood services and neighbourly feelings, Sullivan says. They felt like they'd been taken to the cleaners by the likes of Dr. Charles Allard, an Edmonton entrepreneur widely suspected of using his Socred connections to buy up land at pre-boom prices. The community needed a union, Sullivan argued, to counterbalance the domination of a company town.

The community orientation "lent itself to a political-style campaign," Sullivan says. He conducted a door-to-door canvass, talked to family members, and got supporters to put signs on their lawns. Sullivan and Woodward postered the town and hung banners across stands of tall trees. But if they were all ready to paint the town red, they miscalculated. In November the union lost the vote 151 to 138. Woodward, a Metis, went on to form the Metis Association of Alberta.

Sullivan lost out again a year later, when he tried to catch the company off guard with a hard and fast campaign. He held a community meeting in April 1969 to discuss a union housing policy. "It was well received by the women but there is some doubt as to the men's reaction," he wrote Reimer. His leaflets stressed the need for an international union with professional bargainers who could stand up to an international company. The company-sponsored employee association countered with slick newspaper ads. We are "frankly, a selfish organization," the ads boasted, and we won't be caught playing "second fiddle" to the needs of other units in an American union that locks all its units into cross-Canada bargaining. The association won the April vote, 186 to 103.

Still the company came out second best in the campaign. To defeat the union the association started to play down its links to the company and play up its commitment to fight hard for local interests. In the process it lost its value to the company as a security blanket against unionism. In 1969 the association went on strike against the Great Canadian Oil Sands. In 1971, after defeating OCAW yet again, it launched the first of 65 legal suits protesting rental policies in company houses. In 1975 it christened itself the McMurray Independent Oil Workers Union. "We made our own decisions, kept closer together than we could be in OCAW," says Don Marchand, MIOW president for most of the 1970s. "Our money stayed in Canada. That was the big thing." The local affiliated with the Council of Canadian Unions, a small labour federation opposed to U.S. domination of Canadian unions.

The hard union edge of the local reflected the new workforce that came to Fort McMurray in the 1970s. About half of the new workers were recruited from Cape Breton and Newfoundland after closures at the Come-by-Chance refinery and Labrador iron mines — both unionized operations.

But the local's militancy also reflected a total breakdown of relations with the employer. The old habits of "sweetheart agreements" — winks, nods, and bonuses dished out by the company to reward workers for staying non-union — went by the boards. In 1977 outside labourers wildcatted simply because the company was following the collective agreement on overtime rather than its previous lax practices. In 1979, when the company volunteered a free dental plan, the local almost turned it down. The company's up to some trick, many thought. That's a phase employee associations go through, Reimer says. "The reason why they get so mad and emotional is that the company lets them down, after they do what the company wants them to do," he says. "As soon as the honeymoon's over, the pendulum swings and they decide to scare the hell out of the company. Then we, who were once too militant, become too respectable."

By 1978 the company (now called Suncor) was exasperated with its wayward offspring. "You can't be any worse than what the hell we've got," a senior Sun executive told OCAW international president Al Grospiron. When oil prices collapsed in the 1980s, company paternalism was brought to a rude end. A senior Suncor personnel officer later confided that the company had two choices. It could coddle the workers and resolve their grievances, he said. Or it could take the union to the wall. It took the second option. Supervisors were told to run their departments as they pleased. The company went about denying grievances so routinely that it developed a form letter to cover all situations. In 1986, says Dan Comrie, a former leader of the MIOW and later president of the ECW local, the union filed over four hundred grievances.

"It was really a war," says Sheila Greckol, the Edmonton lawyer who took the local's cases. "It was like nobody was home at the shop. The overtime was incredible. That led to drug abuse and ravaged families. There was a lot of theft, vandalism, time-wasting. Employees just weren't working like they would for an employer they respected." The local didn't give an inch, nor did the company. "Bringing a lawyer into those situations is at best a bandaid," Greckol says. "You need to resolve those problems on the shop-floor. Lawyers just bastardize the process." She told Don Marchand he needed the expertise of a larger union.

To put the organizing drive on a full-time footing, Reimer assigned Ian Thorn to Fort McMurray in 1983, with instructions to show the flag by developing a community base. Thorn had the time for community work. The union didn't represent a single person among the four thousand potential members in the oil-sands industry. Thorn quickly set up a labour council for all the local unions in town. As its first project the council spruced up the town's main park by restoring an historic pioneer home built by an early bush pilot. "That was one of the first steps in saying, `Hey, we're all one big community of labour,'" Thorn says. By 1985 he was president of the town's labour council and the local United Way.

"We knew we had to be there, so that when the time was ripe for workers to go to a union it would happen," Thorn says. "The first step was to gain an image, to show what kind of organization we were. We had to get that acceptance. Being in good stead with the community made the Suncor local confident to contact us in 1986."

From the fall of 1985 on, the town was braced for a strike at Suncor. The company was hard hit by the tumble of world oil prices. Its costs for recovering oil were the highest in the country — the sands are hauled by 6,600-tonne drag-lines to conveyor belts, then put through the wash at 482 degrees Celsius to steam out the oil, which is piped to Edmonton — and executives said the company wouldn't survive without a major austerity program. Managers also wanted to give the local union a licking. The local thought the company still had some elbow room. The province was charging it bargain-basement royalties, knocked all the way down from 12 to 1 per cent to help it through the recession. As the bargaining deadline approached, Marchand suggested a last-minute compromise: Leave the agreement as it is, and we'll negotiate raises when prices go back up. "You're too fucking late," he was told. On May 1, 1986, the company served the union with notice of lockout. The union served the company notice of strike the same day. It also announced its affiliation agreement with the ECW, effective that day.

Suncor made moves to smash the union. Picket lines weren't important to the outcome of the strike, as they are in less-skilled industries where mass picket lines can block scabs from being recruited off the street. The company sent in retired supervisors by helicopter from other locations and bunked them in with local supervisors for the duration. The union had better things to do than stand by plant gates watching supervisors work. Yet within a week, the company sought two injunctions to limit and eliminate pickets. It also sued the union for $5 million dollars in damages.

143

The union said it would obey the May 6 injunction limiting the number of pickets to 20 per gate. But on May 8, the day of a provincial election, an estimated two hundred police were at the plant, armed with full riot gear and fifty guard dogs at the ready. Word got back to town by jungle telegraph. "There was total shock that we were not only fighting the company, the government, and the courts, but also police from across the province," Thorn says, "and our members said `no fucking way.'" Police advanced in pairs to arrest and handcuff each picket. "And as each group was taken away, they were replaced by their wives or other members. It was just something that the women did themselves."

For the duration of the strike, 20 pickets stood at each gate, surrounded by police, dodging when company buses came at them full speed. Any extra strikers had to stand in the ditch. They still managed to take part. "The guys became so sophisticated," Thorn says, "that they could take every window, every piece of glass out, headlights, taillights, the whole works out of a bus, with 16 policemen around there, and not a policeman could identify who did it." This was best done around three in the morning, local activist Keith Barrington says, when police flown in from out of town were dozing in their patrol cars. In the course of the strike, there were 152 arrests from among the 1,050 families on strike.

The aggressive police methods outraged the local community. "I thought I was back in Poland," a local Catholic priest said. Thorn's guess was that the officers were getting "on the job training" for the Olympics and a chance to strut their stuff for an anti-union political show. The scare tactics backfired, he says. They forgot, Thorn says, that "This is a very close community and these were all foreigners, these RCMP officers that came up." Even local businesses supported the strike, Dan Comrie says, because they saw that the union had offered a compromise and the scabs weren't spending a cent in town — they were being flown to Edmonton on their days off.

Bargaining was stalemated throughout the spring and summer as the company continued to toy with the possibility of breaking the union. The company sued the union for unfair bargaining, claiming the union hadn't fairly presented its last offer to the workers. The labour board ruled in September that the union was innocent of any unfair labour practice but guilty of misrepresenting the company's offer. Basken, working alongside national bargaining specialist Buck Philp, suspected the company was stalling, waiting for the union's strike fund to be exhausted and for strikers to start a back-to-work movement. But when an anti-strike meeting was held in early October, it attracted only 25 people.

Basken had been meeting off and on with Suncor executive vice-president Mike Supple since the summer, trying to break the deadlock. Basken did up a sketch of Supple's three choices: break the union altogether; cut a deal with the independent union and cut out ECW; or recognize and trust ECW. Basken warned that ECW was prepared to fight with all its resources if the company chose option one or two. He also supplied Supple with names of people who would vouch for the union's integrity. He and Supple "just had an awful lot of discussion," Basken says. Until the failed back-to-work meeting in October, Basken didn't know which option the company would take. After that meeting he was invited to have breakfast with Supple. They went on to work out a deal in less than two weeks.

The final package, as with most first-agreement strikes, was a saw-off. The company won relaxed language on workplace practices to boost productivity. The union blocked wage concessions, and got an understanding that wages would be reviewed if the price of oil went up or the cost of labour relations went down. Supple immediately granted an interview with the Fort McMurray *Express* to announce his change of heart on union relations. Now, he said, "We can start addressing the real problems of the business jointly instead of dealing with all kinds of bureaucratic and political games." He said, "The major change to my mind has been the intervention of the ECWU that brings a lot more know-how and skill to the table."

These warm words from a senior executive might have been taken as the kiss of death. But the new relationship delivered. Shortly after the strike, union and management co-sponsored an employee assistance program for workers with personal problems that interfered with their jobs. Both sides worked to resolve grievances directly, without resort to arbitration, an expensive process that easily eats up ten thousand dollars a day in lawyers' and arbitrators' fees and lost time for witnesses. As a result of labour relations savings, bonuses and wage increases were granted mid-term in the contract. The next contract, adopted in 1990, brought the company closer into line with national rates.

At the onset of the strike the independent union had signed on for a two-year trial affiliation with ECW. At that time, activist Keith Barrington says, workers saw ECW "as just one large strike fund. But towards the end, people started to reflect on the support and the professional way they helped in bargaining." The Suncor local formally merged with ECW in 1988.

The affiliation procedure is unique to ECW. Reimer provided for it at the ECW founding convention, based on his frustrating experience trying to organize locals like Suncor. "It's the most successful organizing technique we've ever tried," he says. "Instead of making a frontal attack, coming from the outside and trading insults with insiders, we can offer assistance and treat these organizations with the respect that goes along with a merger. There's no loss of face. And our supporters don't have to sit on their hands waiting two or three years for the next organizing campaign. They can get elected to their local executives and fight for affiliation. If the local doesn't like it, it can always back out. It's like shacking up before marriage." The tactic has been so successful, Reimer says, that the oil industry may give up sponsoring independent associations.

145

Solidarity & independence

THE ISSUE
IS JOINED

Rome wasn't built in a day. Canada wasn't built at all. It just grew, "from sea unto sea." That's why our Canadian renovators can't find the blueprints, or why they keep coming up with ones that don't square with the facts. It's the same with unions. They didn't start off organized. Workers just joined them, then had to reorganize them later. That's a problem people face when they muddle their way through history.

Until the 1920s, tramps were the mainstay of unionism in Canada. In the course of their travels they brought U.S. unions to Canada. Highly skilled "tramp artisans" were the country's first long-distance commuters. When there was no work in Hamilton or Toronto, printers and moulders left for

Detroit or Buffalo. When there was no work in Halifax or Saint John, carpenters left for Boston. When there was no work for cigarmakers or textile workers in Quebec, they jumped the border to Massachusetts. When the mines laid off workers in Alberta and British Columbia, packsack miners left for Colorado or Montana. The labour market was continental, and a union card — or travelling card, as it was often called — was the only passport workers needed. Don't leave home without one, said the international craft unions that enforced union-only hiring policies.

Canadian factory workers looked to U.S. industrial unions during the 1940s. Unionists to the south owned the legend for taking on the bosses in the 1930s. Canada's early

industrial unionists weren't colonials. They just went with what worked. They mostly thought of themselves as internationalists, but they mostly confused that with imports. They didn't think through the unique problems of building a labour movement in a cold climate.

Canadian workers could unionize without a blueprint, but they couldn't organize. They imported U.S. jurisdictions in a country with a branch-plant and resource economy, with one-tenth the population, three times the travel problems, and four times the snow. In the 1950s it cost OCAW 20 cents to service one member in Texas, and $1.26 to service one member in Canada.

In Sweden labour solved that problem by pooling and concentrating

TO AMERICA WITH LOVE

resources among 50 unions, each averaging more than 40,000 members. Not the Canadians. In 1956, the year the CLC was founded, 27 of Canada's 178 unions had less than 500 members, and 64 had less than 5,000 members. Only 16 unions had more than 20,000 members, then considered the break-even point for proper levels of service. In 1988, 89 of Canada's 287 unions represented less than 1,000 workers. Only 30 unions had over 30,000 members, then considered the new break-even point.

Without pooling and rationalization, large numbers of Canadian unions are structured into low levels of service or permanent dependency on a U.S. "parent." That's how the battle for increased Canadian autonomy and mergers was joined.

Jim Breaux and Pat Friloux should have been raised in Nova Scotia, where their ancestors settled in the seventeenth century in the French colony of Acadia. But they were raised in the back-country of Louisiana, where Acadian exiles fled after the Yankee invasion of the mid-eighteenth century and became "Cajuns." They returned to a hero's welcome in Canada in 1978, as leaders of a three-year-long Louisiana strike that had been adopted for support by the Canadian Council of the Oil, Chemical and Atomic Workers. The U.S. union's strike against Cyanamid had dragged on so long that Neil Reimer thought it would go under unless someone breathed fresh life into it.

When Breaux and Friloux described their ordeal with Deep South labour relations at the OCAW's 1978 conference in Sarnia, there wasn't a dry eye in the house. Canadian Council delegates raised $10,000 for them on the spot. It was the first outside support they'd received. "It's going to be hard for me to make them believe how true you guys are to the labour movement and to helping friends out," Breaux and Friloux said. The donation saw them through another six months until they won their strike.

With Breaux and Friloux looking on, the Canadians got down to the main business of their conference, which was completing the decision to break ties with their U.S.-based International and link up with two Canadian unions that had broken away from their internationals. A year later, Canadian members of OCAW won support from their International convention for full independence and a future limited partnership between equals. Americans don't make those decisions often or easily. But the Cajuns backed the Canadian decision to the hilt. "I trust them and love them and they love us, " a Cyanamid strike veteran said in a National Public Radio interview. "We went up to Canada in 1978 when we were down in the trenches for 45 months, and they came to our rescue down in Louisiana and they helped us. So anything the people in Canada need, I'll be on their side."

In April 1980, the former members of three international unions of energy, chemical, and textile workers merged to form the Energy and Chemical Workers Union. Although several unions broke away from their U.S. internationals and several unions merged during the 1970s and 1980s, the process that gave birth to the Energy and Chemical Workers Union remains unique. It's the only union that achieved independence and merger in one fell swoop. It's the only union that combined a strong sense of Canadian identity with clear recognition of Quebec's right to autonomy, verging on a relationship of "sovereignty-association." It's the only union that achieved independence after a prolonged period of dialogue within its International and left with the best of mutual feelings.

As Reimer saw it, the new union wasn't a nationalist exercize. "I don't know if I ever considered myself a nationalist," he says. "I consider myself an internationalist first, but you can't become an internationalist by depending on another country to tell you how to practise it." On a practical level, Reimer backed an independent union because it was the only way to unify Canadian energy, chemical, and allied workers into one union. On a more spiritual level, reflecting his Mennonite upbringing, he counted on independence to build character, to foster personal empowerment, self-control, and responsibility. "The search for identity and dignity is also manifested in the development of trade unionism," he told a 1971 Canadian Council meeting favouring Canadian autonomy within the International. That's why he made a point of toning down anti-Americanism. "It's easier to separate when you hate each other," he says. "Hate is a strong motivator, but so is respect and concern for another person's integrity. It takes no courage to drape yourself in a flag, but it's the quality of life and the nature of human relationships we have to address."

When Reimer became Canadian director of the Oil Workers in 1954 he was joining the low end of the international jet set. There was no direct flight from Edmonton to Denver, so he hopped onto milk-run flights whenever he had to go to International headquarters. He rode the old two-engine DC-3, the Model T of the air. He says, "They didn't fly much higher than the mountains. We'd be going north to south, and the wind would come through those gullies and just blow the plane around."

It was always hard making connections between Canadian and U.S. sections of the union. Arrangements were makeshift, budgets were low, headwinds were strong, and knuckles were white most of the time. But that's how unions do business. Though tension and conflict between leaders in both countries were long-standing, there was enough goodwill to work through the basic principles of Canadian self-rule within an international. Until 1975, when the issues of Canadian autonomy were overtaken by other controversies, the union had every reason to think it would stay together.

When the Oil Workers International Union expanded into Canada in 1948, the U.S. unionists considered themselves the big brothers of the International. They had baptised themselves with the world-class International title a good 15 years before they represented one local outside their own borders. They understood their own worldly interests. Oil Worker president Jack Knight promoted expansion into Canada as a move to protect U.S. pay-rates from cheap competition. "How can we maintain our standards if you have a Mexico in the South and a Mexico in the North of us?" he asked U.S. delegates who thought the union

151

should look after its own. When Canadian oil workers joined the International in 1948, the Americans promised to cover the cost of seven Canadian organizers, thus providing a staff that seven hundred members of a Canadian union could never have financed.

But Canadian oil workers never fell into a dependency relationship and never had to break out of one. In the 1940s, individual locals organized on their own steam, acquiring their charters from the Canadian Congress of Labour. They weren't the brainchilds of a U.S. organizer who shepherded them directly into an international. They weren't recruited to an international, they joined one, and they made it clear that they expected to be treated as foreign nationals, not just another organizing district. Canada was recognized as a distinct district — not separated, as were the U.S. districts, along geographical lines.

At any rate, the International counted on the Canadians being self-reliant. After the disastrous California strike of 1948 the U.S. union couldn't afford to hold its convention, let alone pamper Canadians. Though the Americans recovered, and subsidized Canadian losses ranging from $50,000 to $100,000 per year during the 1960s, they watched every penny.

Apart from benign neglect, the Canadian fact gained force from the Oil Workers' unusually anti-authoritarian, democratic structure. Knight rose to leadership in the 1940s as a rebel out to build a member-run, not a staff-run, union. An executive board made up of working members acted as a permanent check on the power of the president. The president had to win its support on an ongoing basis, not just at election time. And because the board was small, the president had to rely on hard-edged deals, not rhetoric, to sway votes. Larger boards may give the appearance of broader rank and file input, but in fact they are easier to manipulate, turn into a sounding board, rubber stamp, organizing vehicle to take the president's line to the members, not the members' concerns to the president. Presidents prefer them to small boards that can develop their own team spirit, follow the administrative details that define an organization's style, take on the top officer without fear of a major blowout.

On small boards, minorities have a stronger voice if they can play off divisions within the majority. Bill Ingram, an early Canadian member of the International executive board, was last but not least in any voting lineup. As representative of the union's newest district, he voted last in all roll calls. He says he was courted as the potential tie-breaker in any dispute. Canadians were also admired for their lively style of "social unionism." Alex McAuslane and John Kane always stole the show at conventions with their recitals of old-country labour lore and songs. Fred Theobald was maestro of the harmonica, Cy Palmer robbed them blind in poker. Reg Basken and Reimer wowed them in debate. As a rule of thumb, Americans are better at delivering speeches than at stand-up debate, because of the presidential as distinct from parliamentary style of political discourse. Canadian labour's involvement in politics gives more workers a chance to speak. "Plus, we could drink them under the table," Reimer says. "We mesmerized them with our spirit."

Apart from their distinctive presence, Canadian leaders demanded respect for their autonomy. At staff meetings in the early 1950s, McAuslane kept his nose out of the U.S. directors' business and assumed they would keep their noses out of his.

"There was the Canadian problem and there was the U.S. problem," says former California director Charlie Armin. "It was very difficult for us to merge the two things into a common discussion" that was truly international, he says. "It never really cleared itself up."

After he became director in 1954 Reimer set up the Canadian Council. "Authority is what you have, and what you take," he says, the organizational equivalent of body-building that goes for the burn. "We never asked permission." He guarded his turf jealously. In 1955, a U.S. International vice-president inspected a strike in Sarnia. The next day Reimer called Knight. Get him out in 24 hours or I quit, he said. "I had no nationalism in mind, but if I'm the director, I'm the director," he says.

Once, when the International's organizing director sent instructions to Reimer, the American received in turn a business-like memo. "Your letter has been received, its contents duly noted and filed," it said. "You're telling me to fuck off," the organizing director screamed back via the telephone. Reimer practised his injured tone of voice. "I never said that," he said.

At International conventions Reimer led a disciplined caucus that delivered a bloc vote. The Americans had never seen anything like it. They took pride in voting as individuals, in line with their training in a political system that puts a premium on the voting record of individual Republicans and Democrats. Canadians came with political instincts honed by parliamentary parties and decades of experience using tight minority groups to hold a balance of power. In a close fight U.S. votes tended to cancel each other and the Canadian vote would tip the balance. "How else are you going to count as a small fish in a big pond," Reimer told his delegates.

In 1955 the creation of the Oil, Chemical and Atomic Workers allowed relations between Canadians and Americans to be worked out afresh. Reimer and Audrey Bruyea, the former Canadian directors of the now merged unions, agreed in a joint statement of February 23 that their objective was "to give the Canadian members a large measure of autonomy on political and economic matters."

Bob Kirk, Canada's representative on the International board, was referred to as "the cold wind from the north." In 1955 Kirk complained to Knight about the waste of mailing U.S. political action material to Canadian members. This could only "emphasize the lack of this type of assistance to the Canadian membership," he wrote. Kirk wanted research, bargaining, publicity, and politics to be made-in-Canada. In 1958 he promoted a constitutional amendment that recognized Canada as a distinct society with more distances, language groups, and variations in labour and social legislation than any U.S. district. Canada needed its own administration, which "may vary with the structure that may prevail in the United States districts." When the amendment passed, Reimer got a raise, two assistants, and became the most powerful director in the International.

Finances were a major source of strain throughout the 1950s and 1960s. OCAW had been based on a false promise. When the Oil Workers merged with Gas Coke, they counted on a $50,000 dowry from the CIO and a $50,000 loan from the Autoworkers to make up for the low dues that Gas Coke insisted on. The UAW never came through, but Gas Coke members still wanted a cheap union. They even opposed a strike fund. "So now, instead of two broke unions, we had one," Reimer says.

When the 1957 International convention failed to solve its financial crunch, Reimer raised Canadian dues on his own. "If we wait," he told the Canadian Council in 1957, "many of our opportunities will disappear." He asked for an extra 50 cents per member per month to fund organizing and expanded services. Why 50 cents? a delegate asked. "It should be a dollar, but I thought 50 cents was all I could get out of you cheap bastards," Reimer said. The motion passed. Five years later, Reimer set another levy on Canadian members, a one-shot contribution of two dollars per member to cover high operating costs in Canada.

The Canadians had aggressive projects in mind and couldn't cope with the Americans' compulsive bookkeeping and penny-pinching. Obsession with economizing turned the 1962 convention into a "complete flop," Reimer told a confidential staff meeting in April 1962. If the same thing happens at the 1963 convention, there'll be an upheaval, he said.

Senior Canadian union staff conducted running battles with U.S. leaders-turned-accountants. In 1960 McAuslane tried to outdo two other unions in an organizing drive but he couldn't buy basic supplies without authorization from afar. The organization is "so cramped up with red tape that it immobilizes action," he wrote.

In 1962 Reimer complained to Knight that the International's strike-pay forms reminded him of a "school teacher/student approach." Knight found two of Reimer's letters that year too "rude" to answer.

Until 1963, Canadian autonomy was defined in administrative terms. Knight politicized the issue by doling out free political advice to the Canadian public. Events of that year explain why differences over foreign policy became so central to the struggle for Canadian autonomy.

In February Knight talked about nuclear weapons to an OCAW conference in Regina. He couldn't have picked a more sensitive issue, or a more sensitive time -- just before a federal election dealing with just that issue. The NDP, backed by Canadian unions, opposed nuclear "tips" on "defensive" missiles placed in Canadian silos. The U.S. government pressed hard for fully loaded missiles in Canada, but the idea of turning Canada into a target for a Soviet first strike didn't seem like a fair share of the cold war. The debate tortured and split the Conservatives, the Liberals, the entire Canadian public.

On the eve of the conference Knight held a press conference to announce that a rifle wasn't much good without bullets. The next day the Saskatchewan media were awash with news of the U.S. labour leader who had put Canadian radicals in their place. Both Reimer and Reg Basken, president of the Saskatchewan Power local hosting the conference, worked Knight over, then issued a press release dissociating themselves from Knight's views. "We resent an American labour leader coming to Canada to pressure us into accepting policies of the Pentagon and U.S. State Department," Vancouver leaders of OCAW told radio station CKLG. In a national press release the West-coast unionists said that Knight's "unwarranted interference in Canadian labour policies points to the need for Canadian autonomy, that is, the right of the Canadian membership to decide on its own policies within the international."

The unravelling of cold war politics may have been behind the clash between Reimer and Knight at the August 1963 OCAW International convention in Chicago.

On the eve of the convention, Knight summoned 40 union leaders and gave out the marching orders for the week. The big controversy was over a motion to make every local pay into a strike fund, a motion that Knight opposed. Above all, Knight wanted to head off a bitter fight on the convention floor by getting an initial show of support from the 40 leaders. He got 39 votes. Reimer, seated to his immediate left, spoke last. You won't get one vote out of Canada, he said. Canadian delegates wanted a big war-chest. Knight told Reimer to get in line. "Sorry, you're telling the strong to go to the weak," Reimer said. "I refuse to make myself a whore."

Word of Reimer's defiance hit the cross-border grapevine immediately. The next night a meeting was held to draft him for a run at the International presidency. Someone came with buttons: "It's been Knight too long." Reimer says now he was "a bit excited" at the time. But when he checked with his own district, two members spoke strongly against him as president. That knocked the wind out of him, killed his possibilities for a strong start. Reimer later found out that

McAuslane had been promised the directorship if he could split the Canadian caucus.

All next day, Reimer felt eerie. He saw people he'd never seen before, wearing suits that didn't fit with oil and chemical workers. A senior head office staffer tipped Reimer off that the phone in his hotel room was tapped. Convention hoopla got very tense. The halls were abuzz with rumours that socialists were taking over the union. "You all come from a socialist country," one U.S. delegate told John Kane, a delegate from the Clarkson local. "Even your garbage is collected by the government."

On Wednesday morning Knight came to Reimer's room for breakfast. Although Reimer had pulled out of the presidential race, Knight wanted a deal on the strike-fund debate. Within minutes Knight threw his bacon and eggs on the floor and stormed out of the room. Reimer's life was falling apart. Until then, he'd been Knight's favourite, was seen as his protégé. He was shaken by the opposition in his own district. He had pangs in his own conscience, hard for anyone outside the labour movement to understand, about betraying the union leadership blood oath of "solidarity in the ranks, unity in the leadership." Breaking this code of executive solidarity, showing "weakness" and "division" to the enemy, is akin to treachery for most union leaders. It's pointless to argue that members aren't the enemy, that they deserve articulate and informed champions of dissenting views, that solidarity isn't the same as uniformity, that labour's loyalty oath stifles new ideas. Unions are combat organizations, not debating clubs. Reimer gave in to his guilt, found Knight on the convention floor, and offered his resignation as director. He left the convention and flew to Toronto for a few days on his own.

For the rest of the week, Basken, the Canadian representative on the constitution committee, led the fight on the floor for a compulsory strike fund. A union without a war-chest was a company union, militants said. The rebels won a strong majority on Wednesday but not the two-thirds required for a constitutional change. Once the split was out in the open, Knight moved behind a compromise resolution that carried on Friday. Before the convention broke up, Knight named McAuslane Canadian director. The Canadian caucus refused to accept him. Knight then chose Ron Duncan, a former workmate of Reimer's at the Co-op Refinery in Regina. Duncan was the newest member of the Canadian staff and had spent most of his probation period in hospital, recovering from a car accident. Though Duncan was bright and full of good ideas, Knight probably chose him because he'd had the least time to make enemies for himself. Duncan immediately demanded that McAuslane retire.

Reimer spent those days at the Royal York in Toronto, nursing his ego, sorting out his life and his values. He expected the worst from Knight, who was known to insiders as a widow spider — widow spiders kill their mates. To protect Buck Philp and Maurice Vassart, his closest allies on staff, he called Tom Sloan, Canadian director of the International Chemical Workers, and said Sloan should consider picking up the two oil workers if they got fired. Sloan, apparently enchanted with the idea of a bloodletting in OCAW, phoned McAuslane, who informed Knight. When Reimer got home, Knight's telegram was waiting for him. He'd been fired for disloyalty.

Reimer appealed the firing and was duly summoned to a hearing in Knight's office. Knight had a long boxcar-style office where he usually kept his desk and a set of chairs close to the door so visitors didn't feel too distant when they popped their heads in. On this occasion Knight had moved his desk to the far end, so Reimer could get the full feel of walking the corridors of power. "It was the classic American show of power," says Reimer. "You come to me. It was just like in the movies. He sat firmly behind his desk, not there to greet me at all, and I had to walk 40 feet to his desk."

Reimer gave his side of the story, that he was just looking out for his most loyal staff. Knight wasn't interested. He changed the subject. You can go back to being director, he told Reimer, if you confess that Charlie Armin put you up to it. Armin was Knight's executive assistant and one of Reimer's best friends. Reimer left, passing by Armin's secretary, who noted that the Canadian was "as white as a sheet." Knight did hire Reimer back, but he demoted him, put him on indefinite probation, and ordered him to stay out of union politics.

Knight had, in fact, lost his grip on reality. His photographic memory had once been his pride and joy. Members loved how he remembered their names, what they'd said ten years before. When his memory snapped in the 1960s, he couldn't come to terms with it. He pretended nothing had changed, but he became victim to every con artist who "reminded" him of what he'd agreed to. "It is pretty spooky here, what with all the empty rooms and nothing but occasional confusion to interrupt the silence," public relations director Ray Davidson wrote Reimer from head office in November 1963. "The man is very very ill, grayness more remarkable, eyes troubled, never a smile any more under any circumstances -- just tired look or angry. Hands shake, voice fades. It is awfully sad to watch." Modern medicine identifies this disorientation, memory loss, and vindictiveness to intimates as the classic symptoms of Alzheimer's Disease.

In 1965 Knight asked Reimer to line up Canadian support for Knight's chosen successor. Reimer reminded Knight that he was still on probation and couldn't get involved in internal politics. A dissident, Al Grospiron, beat out Knight's man, thanks to solid Canadian backing. "So I had the last laugh," says Reimer. Grospiron reinstated Reimer and reappointed him director in 1967. Grospiron never forgot what he owed Canada.

Reimer's recall of the 1963 convention incidents remains blurry to this day, an exception to his own near-photographic memory, a sign that he still blocks it out. "It was personally shattering for me," he says. The experience changed him in two ways, he says. "People don't get much of a second chance with me." When McAuslane died several years later, Reimer refused to attend the funeral. "I buried him long ago," he told those who wanted an explanation.

But he disciplined himself to give the International a second chance. He'd seen too many union activists fall victim to their disappointments, centre their lives on settling old scores. He didn't want to rebuild his career on a platform of getting even. So he kept word of his firing and demotion to himself. "It would have been the death of the International in Canada," he says. He told supporters that his fight was personal between him and Knight and didn't involve the union. "You can't build a union by disloyal acts," he says. "The first lesson is you have to be proud of your union, and proud of the fact that you can resolve differences internally, not run away because you don't think you can resolve them."

Ironically, his treatment at the 1963 convention never became evidence for why Canadians had to go on their own to win true independence. Instead, his rebound from that experience explains why Canadians left their International with the best of feelings towards it. "I never wanted the debate to centre on what was wrong with OCAW," Reimer says, "but what we could do on our own."

When Reimer and Charlie Armin met in the late-1980s for an interview about the 1963 convention, Armin hinted that Reimer's supporters in the presidential run set him up for a fall. Armin told Reimer, "This thing had a lot more fingers reaching into it." Even the CIA was involved, as the Denver *Post* and *Ramparts* magazine revealed in 1967. Since 1961, Knight, president and chief financial backer of the Denver-based International Federation of Petroleum and Chemical Workers, a world-wide coalition, had been the conduit for CIA funds laundered through the Andrew Hamilton Foundation.

Reimer had asked Knight where he got the money to finance the Federation. He knew the debit in OCAW's account was too small to cover it. Knight brushed Reimer off, said the money came from an anonymous wealthy widow who believed in liberal causes. "Well, with Jack, that was sort of a credible story," says Reimer. Knight cut a dashing figure and "had the ability to know those kinds of widows."

The U.S. government had long used oil as a penetrating fluid to ease into the inside workings of the world labour movement. In the late 1940s, when the U.S. Marshall Plan funded the rebuilding of war-torn Europe, the U.S. government leaned on the coal-rich nations of Europe to import oil. The U.S. government discouraged European reliance on their rich coal reserves, according to Carl Solberg's *Oil Power: The Rise and Imminent Fall of an American Empire*, because they suspected coalminer unions of Communist leanings. In the 1950s the oil industry provided a strategic opening for CIA influence because it operated in so many countries and refinery workers were key players in Third World labour politics. As well as leading OCAW and the International Federation, Knight held down the Latin America desk for the International Confederation of Free Trade Unions, set up by U.S. unions at the onset of the cold war to challenge "pro-Communist" unions. Knight was a bird dog for President Eisenhower and Kennedy on all their excursions to South America. He made sure they met the right people and knew all about the wrong people who mobilized mass anti-Yankee demonstrations.

There's no proof that CIA operatives influenced Knight's treatment of Reimer in 1963. Nor is there any evidence of RCMP tampering, though the RCMP had three informants among the union's secondary leadership, according to off-the-record interviews conducted for this book. What isn't open to dispute is the importance of world politics to the evolution of U.S. unions. That's why the leftward evolution of Canadian labour during the late 1960s pushed union foreign policy to the forefront of demands for autonomy. In OCAW, that controversy might have been even sharper, except that Reimer deliberately downplayed ideological differences as he moved to put the pieces in place for a smooth transition to a separate operation. For Reimer the Canadian operation had a strong enough case for separation on pragmatic grounds: "It had to be done for us to survive." To the extent he succeeded, questions about the events of 1963 vanished into history.

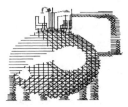

From 1963 to 1974, the battle for Canadian autonomy within OCAW centred on the rights of Canadian members to make their own financial and policy decisions.

Though Duncan owed his appointment to Knight, he was no patsy. He blew his top when Reimer told him that the Canadian director had power to sign for only up to $35 in expenses. In 1964 Duncan was denied permission to charge the $75 registration fee for a labour-management conference to the union. It's impossible to make long-range plans under this regime, Duncan protested.

When Reimer returned as director in 1967 he kept up the pressure. In 1970 he had a temper tantrum when an International research officer refused to share information on world energy trends until after President Nixon announced his oil policy. Later that year Al Grospiron refused to pay for Canadian office equipment and warned Reimer that Canada was getting "a little expensive." Reimer asked for "any suggestions that you may have that would shrink the size of Canada." Reimer suggested that Grospiron look elsewhere for cutbacks. "Most of the `Service

Department' of our headquarters are relatively useless to this district," he wrote.

Reimer beavered away to keep Grospiron up to date on Canada. In 1970, for instance, he sent the president an *Edmonton Journal* item on Canadian nationalism, along with a personal note saying that "We have to lay plans to stay ahead of this so-called `nationalistic' feeling that is emerging in Canada." He warned that international union offices were too slow at giving Canadian autonomy. Unknown to the president, Reimer helped to ghostwrite the speech Grospiron read to the 1970 CLC convention. It was "absolutely essential," Grospiron said, that international unionism not become a "vehicle to impose United States viewpoints on the membership in Canada." Grospiron praised CLC delegates for demanding "self-government" within internationals.

In 1970 and 1971 the CLC set out minimum standards of self-government for Canadian members of internationals. Canadian unions had to elect their own officers, give them power to make their own policies, and speak for their members on all Canadian issues. OCAW's 1971 international convention accepted these guidelines in its constitution. They were necessary, Bill McGough of the resolutions committee argued, "to have a truly International Union, that is, one covering two separate and sovereign nations." As of that date the Canadian director had to be elected in Canada, not appointed by the International president like other district directors.

The 1974 CLC convention seemed like the high point for the nationalist surge around Canadian autonomy. It called for complete freedom of Canadian members of internationals to participate in all aspects of Canadian life, to participate directly in all international conferences, and to set an independent foreign policy. The convention is best known for the coalition of Quebec nationalists and independent public-sector unions that granted far-reaching autonomy to the Quebec Federation of Labour and turned thumbs down on the official slate proposed by CLC leaders. Shirley Carr of the Canadian Union of Public Employees was elected CLC vice-president. Don Montgomery, a Steelworker staffer from Toronto, was the upset winner for secretary-treasurer. He defeated Reimer.

According to Reimer, he was asked to run on the eve of the convention and told he had the unanimous backing of the CLC executive. "I didn't see any philosophical alignments, just saw it as a hell of an opportunity," he says. His campaign alienated supporters. He said CLC staff needed to be brought up to snuff, which didn't endear him to them. He said local labour councils had to be more forceful, which ticked off local leaders, who felt stretched to the limit just keeping their councils alive. Mainly, he ran with the wrong crowd, with leaders who couldn't deliver their delegations, with people seen as stand-patters.

After the convention OCAW was the first union to win autonomy on international questions from an international. In July Reimer convinced the Canadian Council to set up a special international fund to promote exchanges with workers in other countries. In August he charged his trip to meet Soviet oil workers to his International credit card. Grospiron deducted the flight from his salary. You have to ask the International's permission first, Grospiron wrote. "I feel your position places in great question what autonomy we really have," Reimer wrote back. He demanded a formal debate on the matter.

Grospiron called Reimer to a meeting of International officers, where a stack of Reimer's memos would be raised for discussion. Don't give him "a fucking thing," Grospiron had primed his supporters. But in the meeting, as Grospiron worked his way through the pile he ended up agreeing to every request, including covering the expenses to the Soviet Union.

Later, after a few drinks at a local bar, Grospiron confessed he'd finally met his match for stubbornness. "He was a typical Texan," Reimer says. "Fortunately, my first boss was a Texan. So I knew that if I didn't stand up to him, he'd walk all over me; but if I went at him toe to toe, he liked it."

Grospiron did dig in his heels against Reimer's demand to place the 1974 CLC guidelines in the OCAW constitution. Reimer wanted the statement endorsed in principle, and he wanted a separate fund for Canadian expenses at international gatherings. Grospiron, already pushed hard enough by events, didn't want any changes codified. The constitutional change in 1971 "adequately covers" any points about international status, he wrote Reimer in August 1974. Reimer already had a Canadian fund, and the 1974 guidelines added nothing new, so he let the matter drop.

Canadians in OCAW had gone as far as autonomy could take them.

The 1974 CLC convention proved a poor predictor of the issues, alliances, and personalities that would define Canadian union autonomy a few years later.

Until 1974 autonomy was an issue that attracted radical activists interested in policy. After 1974 these concerns took a back seat to more practical problems of union survival, like jurisdiction and bargaining clout. When practical issues became priorities, a lot of reformers lost their nerve and a lot of "standpatters" found their cause. Reimer had foreseen this second set of issues in his 1971 report on the CLC convention. There are two questions, he said in his report to OCAW's Canadian Council. One is autonomy, which we must adapt to. The other is structural. "The relatively small segments of the various internationals cannot afford the luxury of overlapping services and staffs to the degree that this can be tolerated in the United States."

This was the sleeper issue that soon tore the labour movement apart.

THE TOUGH
GOT GOING

Between 1960 and 1990, unions became the fastest growing independence movement in the country. In 1960, 72 per cent of Canadian unionists belonged to U.S.-based international unions. By 1989, as a result of expanding public-sector unions and breakaways from U.S.-based industrial unions, only 32 per cent of Canadian unionists belonged to U.S. internationals.

As long as basic organizing, servicing, and bargaining dominated the union agenda, industrial workers saw no reason to leave their internationals. But as the challenges confronting unionists became more complicated in the 1970s and 1980s, as the United States lost its world economic dominance and ceased to offer stable and continued growth, Canadians discovered they had their own union culture.

Employers huddled closer to the U.S. empire as it lost its world economic leadership. Unions couldn't afford to. They discovered nationalism in the course of discovering their own self-confidence in adversity. When the going got tough, the tough got going — into their own independent unions. Because practical union issues were at the forefront, all breakaways were leadership-driven, not the result of rank and file mutinies.

Each union had its own reason for cutting U.S. ties. Canadian paperworkers made better wages than U.S. ones, mainly because they were willing to strike. U.S. leaders balked at high strike pay for the Canadians and almost invited them to leave.

For autoworkers, "The split and nationalist sentiments were a consequence of defending workers, not goals in themselves," Sam Gindin, leading Canadian Autoworker staffer, writes in the Summer 1989 *Studies in Political Economy*. U.S. leaders not only accepted concessions for themselves, but also insisted that lower paid Canadians take them too. By contrast, bargaining differences played no role in the separation of energy and chemical workers. Mergers were central.

Each international reacted differently. Joe Tonelli, head of the Paperworkers Union, happily gave autonomy but refused to return over $2 million in assets. In 1974, according to John Herling's *Labor Letter*, Tonelli was jailed for embezzling union funds, despite pleas from his lawyer

that President Nixon "got an absolute pardon. To put this man in jail makes a mockery of justice." Tonelli travelled in those circles. He chaired New York's racing commission, represented the United States at the International Labor Organization, and was awarded a knighthood from the Pope. The U.S. Autoworkers leadership refused to give bargaining autonomy — it "was obsessed with protecting its political ass against criticism that it had let Canada go too easily," Bob White says in his autobiography — and did nothing to promote communications once the unions were separate. By contrast, the U.S. leaders of OCAW gave Canadians everything they were owed, maintained a common strike fund, and continued to work with Canadians on health, safety, and environment issues.

"THE PEPSI GENERATION"

It's not so long ago that Blacks were called niggers, and Quebeckers were mocked as white niggers, pea-soupers, and Pepsi's. Cheap protein and cheap pop for cheap labour. In the 1960s, Quebeckers wanted out of the Pepsi generation. They launched a Quiet Revolution to take a taste test as "masters in their own house." They turned to their provincial government to modernize their economy, education system, and social services.

Outside Quebec, the media pointed fingers at separatism. Inside Quebec, the changes were harder to put a finger on. They came, then-Liberal cabinet minister René Lévesque said in 1963, from a people "fed up of being seen as a museum, as the 'quaint old province,' a nation bent on advancing, rising, no longer merely to endure." Lévesque said, "Involved in this is something called nationalism: it is at the bottom merely the very normal desire for dignity. That comes from having control over oneself and one's destiny."

The same spirit turned unions into the independence movement of the workplace. That's where the battle to preserve language and culture in the modern world would be won or lost. That's where a third option — another choice besides being slaves to the past or slaves of modern multinationals — had to be found. Unions became the civil rights movement of Quebec's "white niggers."

As of 1961, John Porter shows in *The Vertical Mosaic*, French Quebeckers were at the bottom of the heap in their own province.

They'd lost ground in the better jobs since 1931 and had made almost no headway out of the lousy ones. French Quebeckers made 64 cents for every dollar made by the English, Raymond Norris and Michael Lanphier calculate in *Three Scales of Inequality*. Almost all immigrant groups were higher on the totem pole. In Montreal, according to Leo Johnson's figures in *Social Stratification: Canada*, the average francophone Montrealer earned $4,113 a year, $1,649 less than anglophones. When money talked, it spoke English. It was the language of management and the better stores. It was the language of machine parts, instruction manuals, union agreements. The price of getting ahead was assimilation. Of young people surveyed for

▼ PART I
▼ SECTION 4 / SOLIDARITY AND INDEPENDENCE
▼ CHAPTER 2

VIVE LA DIFFERENCE

the Royal Commission on Bilingualism, 92 per cent thought bilingualism was necessary to land a job, and 83 per cent thought it was essential for a promotion. "Speak white," the saying went.

Until the government sponsored language legislation, it took strong unions to break down that language barrier, to speak in a language employers understood. That's the pitch long-time union organizer François Gagné always made. It's your only chance to be heard, he said. Your bosses can't even figure out what you're complaining about. "They knew they had to fight as a group to survive. They knew that a lone individual couldn't handle the discrimination."

In the early 1960s, U.S. union leaders were brought up to speed on the hairpin turns of the Quebec labour scene.

Quebec had a reputation for reckless bargaining, and OCAW officials from Denver were insisting on showing the hotheads how it was done. So Canadian director Ron Duncan arranged an object lesson for backseat drivers. He asked staff rep Maurice Vassart to chauffeur them from the Montreal airport to Quebec city. Duncan says, "He was such a good driver, with his hat cocked over his eye and his cigar hanging out of the corner of his mouth, and he had a fairly heavy foot. I suggested he make it a memorable ride. And he did. That old number 2 highway was narrow, and it was winter, so it was icy, with big maple and oak trees along the side." Duncan thinks "they got the message that I wasn't interested in having them intervene in negotiations of that sort, that if we were going to be a viable district, we were going to have to do it on our own."

Quebec workers drove English-speaking union leaders around the bend. Anglo unionists couldn't keep up with the super-charged militancy of the Quebec scene. And, quite apart from language difficulties, they couldn't follow the logic of what Quebec workers wanted from their unions.

After the heady days of the 1930s and 1940s, unions in English-speaking North America settled into a routine of slow and steady progress. Leaders forgot what it had been like when workers took bold initiatives and pushed bargaining to the limits. They lost the passion and vision to put everything on the line. The mood in Quebec in the 1960s was alien to them.

"Things have got to change," Liberals chanted when they swept the Union Nationale party of Maurice Duplessis from power in 1960. René Lévesque, the Liberals' most flamboyant cabinet member, wanted to help provide the "spring cleaning of the century," he wrote in his memoirs. Where was it all leading? "Well, I'm not too sure, but we're certainly getting there fast," he said.

Workers also made up for lost time. By 1968 union membership had tripled, three times the growth rate in Ontario. Days spent on the picket line jumped from 218,000 in 1960 to 1,780,000 in 1967. "After you've been in the ditch for a long time, when you start to emerge, now you're almighty," explains Martial Langevin, OCAW's executive member in Quebec. "Let's take advantage of it. That's the only way — take advantage of it, and go all the way. We are so far behind that if we want to get equal, we have to double up the dose. We've got to work harder, do things faster."

English-speaking union leaders were mostly graduates of the school of hard knocks. They came up from the ranks. They had lots of street-smarts and loads of contempt for academics who didn't. By contrast, meetings of Quebec union leaders tended to be class reunions of former university students of Fathers Georges-Henri Lévesque and Emile Bouvier. In the sullen days of Duplessis, they trained a cadre of secular missionaries in the liberating themes of unionism. Quebec had to be free "from foreign economic control, from capitalist control, and from Church control — all at the same time," Father Lévesque taught. Jean Marchand, Gérard Pelletier, and Marcel Pèpin of the CNTU, Madeleine Parent, and Gérard Rancourt in the textile industry, Jean-Marie Bedard for the woodworkers, Fernand Daoust, Philipe Vaillancourt, and André Thibodeau for the CLC, Claude Merineau in food, Roger Provost at the QFL: all of them went to the head of their class after graduating.

Quebec unions weren't cut off from students, writers, and artists, as unions tended to be on the rest of the continent. Media star René Lévesque, for instance, became a labour and national folk hero in the 1959 strike against CBC, widely seen as a strike for high quality Quebec programming. University students organized a union and struck in 1968. Student solidarity committees helped out on picket lines. Journalists and unions joined hands to build a popular labour weekly. The labour movement was a youthful romance, full of great promises and eternal declarations.

This was the unionism of a "distinct society," a matter beyond the understanding of hardheaded ("squarehead" was the Quebec expression) unionists elsewhere, who were untrained in coalitions and worried about losing control. "I honestly believe that there is a conspiracy of a sort which takes in and has been led by the intellectuals in the province of Quebec," senior OCAW staffer Ed Waddell wrote Reimer after attending a Montreal rally in the late 1960s. Separatists were all over, he wrote, "and what I hate to see but what appears to be quite evident is that there are amongst them some extreme left wingers." As Waddell saw it, "There is absolute madness in that province."

Finally, North American union leaders couldn't keep tabs on what seemed like Quebec's double standard. Until the 1950s, Quebec workers spurned U.S.-based unions as too foreign and materialistic. They opted instead for Quebec-based unions inspired by Catholic doctrines of a separate, stable, and harmonious society. That made English-speaking radicals suspicious about the dark side of Quebec's identity. After the 1960s, when Quebec workers became the most militant and socially conscious on the continent, Quebeckers again rejected U.S.-based unions as too foreign and materialistic. This time they opted for unions inspired by doctrines of a distinct society that needed change through struggle. They wanted their unions to be more than bargaining agents: they wanted unions that met the full range of their social, economic, and cultural needs.

Though the phrase hadn't yet been coined, Quebec workers wanted a labour movement based on "sovereignty association." They needed association — which translates in union terms as solidarity — more than workers on the rest of the continent. They had to catch up with wages paid elsewhere and they had to teach giant U.S. corporations respect for Quebec's language and culture. So they defied union structures based on divisions from another place and time. Quebeckers also needed sovereignty — which translates in union terms as local autonomy — more than workers on the rest of the continent. They needed it to be credible with a people struggling to be "masters in their own house," and to cope with the special needs of a minority culture facing convulsive economic and social change. So they defied union structures based on outside control.

The result of this cultural conflict was constant turmoil. In the oil, chemical, and textile industries, where three unions broke from U.S. internationals to merge as the Energy and Chemical Workers, the result was a great achievement for both bargaining unity and local control. It's one of the few cases in Canadian history where "English" and "French" found common ground based on identical economic needs and parallel needs for democracy and local control, by putting respect for Quebec's distinct society within a framework of Canadian independence.

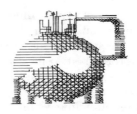

The Oil Workers had only five hundred Quebec members in 1954, less than 15 per cent of the potential members, according to an organizing report.

Massive police repression of the 1952 copper strike, a showcase for Montreal's industrial east-end, was partly responsible for this arrested development. The Oil Workers had greater success at McColl Frontenac (later Texaco), where they won large wage increases by adopting Toronto payrates, but that still didn't serve as the jumping-off point for a province-wide organizing drive. Their problem was isolation from two basic realities of Quebec. Montreal's East End, a centre for refineries, was a long way socially and geographically from the Quebec centres of the chemical industry, such as Shawinigan. Montreal's refineries were also a long way from the French majority. Refineries mostly hired anglophone workers, who supposedly had better technical education and could understand English supervisors and the English-language technical manuals. Refinery jobs were "favoured jobs," says Bill Butcher, then at the Petrofina refinery. "They were lifetime jobs — to get into the oil industry, so to speak, you had it made."

In 1957, shortly after the formation of OCAW, the union got its chance to get off its Montreal island. Maurice Vassart, a partisan with the Belgian resistance during World War II, worked in the underground of the Quebec chemical industry to give OCAW its first contact with Shawinigan, the home base for CIL and Du Pont (which were one company until 1954) and a score of other industries drawn to the area's cheap electric power, plentiful pulp supplies, and conservative customs.

Vassart came to Shawinigan in the 1950s, hired as executive director of Quebec's National Federation of Chemical Workers, a branch of the Confederation of Catholic Workers of Canada. The Federation wanted someone who was bilingual and could talk to the bosses, and its members liked Vassart's dramatic style. A former union official in Belgium and Ottawa, Vassart had problems adjusting to the Federation. He resented being ordered to attend church as a show of good faith. He didn't like the whipping taken by the union when CIL and Du Pont refused to pay standard industrial rates in 1957. A strike "couldn't move anywhere," he says, because the companies just moved production to other operations across the country.

In 1957 Vassart issued a call for an emergency convention of the 13 locals in the National Federation of Chemical Workers. "Before going to war, it is wise to make an inventory of our ammunition," the old military strategist reported, adding that the union suffered from "very severe weakness in our arsenal." The chemical monopolies organize on a world basis so they don't put "all their eggs in the same basket, and if we do not succeed in extending our own operations, our back will be broken very soon," he said. "We will have to elaborate our bargaining strategy and eventually our strike strategy on the national level, and even on the international." Vassart favoured joining OCAW, the smallest but most democratic and "most aggressive and the most militant" of the internationals. Its small numbers guaranteed Quebec workers a decisive role, Vassart said. The convention, representing 4,500 workers, called for a merger with OCAW. The church official who served as pastor for the union warned that he would oppose any switch to an international union.

Reimer's original plan was to sit out while local leaders handled the membership affiliation vote themselves. He hadn't counted on the severe counter-attack. The Confederation of Catholic Workers assigned a top officer to chemical bargaining and sent in 13 organizers to firm up the members. OCAW was denounced from every church pulpit. Vassart was thrown out of public meetings. Insiders couldn't resist the pressure, Reimer wrote the International president in 1958, so OCAW had to shift gears and conduct a traditional "raid." Only two small locals, both based among bilingual office staff, made the switch. Reimer guesses that their bilingualism led them to feel more comfortable in an international union.

OCAW's Quebec operation didn't boom until the 1960s, when its expansion led the country. From 1961 to 1966, OCAW membership went up 1 per cent in Manitoba, 10 per cent in Ontario, and 28 per cent in Quebec. Refinery workers couldn't be stopped. In 1962 Tolhurst Oils fired 20 workers who had signed union cards, and the entire workforce walked out. BP workers, still in the process of organizing, bolstered their picket lines. When Reimer hired Fernand Daoust, a former organizer with the CLC and a vice-president of the Quebec Federation of Labour, Daoust brought along many of the CLC locals he'd serviced with him, and gave OCAW a high profile within the QFL. Quebec locals set up the first area council in the Canadian district, which allowed them to hold their own discussions on provincial issues and develop their own leaders.

Growing pains were inevitable. In Montreal workers were scattered in 35 locals covering 50 workplaces. Only three locals had more than two hundred members. There were only two union staffers to service the members. Relations with employers were rough and ready. In 1964 OCAW stalwarts at Fina brought baseball bats, but no balls, to a confrontation with anti-union managers. The local president at Atlas Asbestos was fired for throwing paint in his foreman's face.

Any breakdown in the tight discipline of pre-union days gave anxiety attacks to plant managers, especially in the chemical industry. The danger of accidents at CIL's explosive plants was so great, says former personnel chief Bob Gallivan, that "Managers had the mindset to do everything by the book. We didn't want workers with any imagination. We wanted employees to follow instructions to the comma."

"There were a lot of bad days and people were under a lot of pressure," says John More, who was president of the BP local in

Montreal in the 1960s. "It was hard to get anything from the oil companies. You either dealt with that, or you blamed somebody," he says. Bill Butcher, sacked by Fina for a year before the union won his job back at the labour board, turned against all international unions. He led a breakaway group, the United Oil Workers of Canada, which recruited a few thousand industrial workers, mostly in the Montreal area.

Reimer pleaded for more funds and understanding from the International. People "are in the throes of complete change, and this is bound to continue for some years to come," he wrote the OCAW president in 1963. Quebec must be serviced in French, he insisted. After 1961, all union agreements in Quebec were negotiated in French. In 1963, *Union News*, the International's monthly newspaper, started covering Canadian events in both languages.

But in 1965, Canadian director Duncan complained that most International mailings were still in English. "This amounts to the same thing as waving a red flag in front of a bull," Duncan wrote the president in Denver. "There is a great upheaval going on in the province of Quebec, and the French-speaking people are demanding that their rights under the constitution of this country are fulfilled."

Fernand Daoust, an ardent Quebec nationalist, kept warning Duncan that all internationals were on the hot seat because of the Confederation of National Trade Unions, the Quebec-based union central made over from the Catholic confederation. "We have to compete with an efficient organization which is taking advantage of this awakening of French Canada and the desire of French Canadians to be fully recognized," Daoust wrote Duncan in 1964, a year of major defections. "Unless the CLC and the International Unions recognize fully this problem, we may have worse news for the future." The CNTU pounced on any locals where service was slack, Daoust wrote in 1967. "The type of servicing job that we are obliged to do due to this situation in Quebec is not comparable with any other place in our Union."

Indeed, intra-union competition made Quebec a province unlike the others and was forcing the pace of change by the mid-1960s. Quebec was the only province where two equal union federations — the CNTU and QFL — slugged it out for members' loyalty. The threat of raids was constant, and unions had to outbid each other with promises of more democracy, service, and results. The CNTU guaranteed Quebec control and service in French. It had dynamic leaders and a large central staff. Thanks to a friendly provincial government that gave it exclusive rights to represent civil servants, it was well funded. It was also eager to liberate workers from U.S.-dominated unions. In 1964 it took ten thousand members away from internationals. The Quebec Federation of Labour watched helplessly, overwhelmed by what it called "raids-a-go-go" conducted by an enemy "panzer division."

The QFL, creature of the international unions and the CLC, was described in Albert Verdoot's study of Quebec unions as a "post office" for the CLC. But it could barely afford stamps. Its budget in 1964, based on eight cents per member per month, was $100,000, about 5 per cent of the CNTU levy. In 1965, according to Emile Boudreau's history of the QFL, the federation called for special Quebec status in the CLC. In 1967 it demanded full bilingual services as well as complete Quebec control of services and organizing. In 1969 it wanted to issue its own charters — a power jealously guarded by individual unions — and pick its own CLC staff. In 1974 Quebec delegates at the CLC convention joined with Canadian-based public-sector unions to win all these changes, along with the right to charter and protect independent locals that had broken from their internationals.

Changing the QFL was the easy part. Small and understaffed international unions couldn't cope with competition from the CNTU and made no moves to overcome their problems. The Ohio-based International Chemical Workers was "not giving good service to the people," says Bernard Boulanger, who fought off CNTU raids for the ICW. In one union vote Boulanger hired an ambulance to bring a man from the operating table to a polling booth, providing a vote that made the difference. Yet Boulanger couldn't convince the ICW to print basic forms in French.

The Steelworkers, largest union in the QFL and already enjoying regional autonomy within its International, put small unions on notice. They'll clobber us themselves if we don't get our act together in time against the CNTU, Duncan wrote the International in 1965. There's no choice but a merger of all unions with a common jurisdiction, he pleaded. "The status quo will only lead us backwards." John Crispo's 1967 study of international unionism made the same point. Small unions can't beat off the nationalist CNTU as long as there are so many of them," he wrote. "Especially in Quebec, there is probably no solution, short of rationalizing the structure of the CLC sector of the labor movement." Union mergers, as well as more responsive services, became matters of sheer survival.

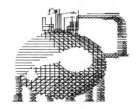

If external pressures made new structures a matter of union survival, internal pressures made them a matter of union rebellion. "The days of quiet unanimity and passivity are over," Daoust wrote Reimer in 1968, after Reimer's return as Canadian director. Quebec was a nation, and should form one of two union districts in Canada, Daoust wrote, adding that he intended to defend that position publicly.

Reimer, who had become immersed in the Alberta NDP after he left as Canadian director, had little inkling of what was going on in

Quebec. He sensed that Duncan had played too much to Daoust, the man of intellect, and not enough to Vassart, the man of action. Reimer suspected Daoust was organizing to split off from the OCAW, and that only a firm hand could salvage anything from the wreckage. That view was shared by Martial Langevin of Shawinigan, who says Daoust urged him to split from the OCAW.

Reimer, back as director, took a hard line at the Canadian Council convention in April 1968. The spirit of our Quebec locals is too inward, he said. He might back an assistant director for Quebec, but only if numbers warranted. "Our Union is an economic organization," he said, so "total and complete emphasis should not be on its social and cultural responsibilities." In response Daoust called an "extraordinary convention" of all Quebec OCAW members for May 18. Before that could happen, however, Daoust resigned to become Quebec director for CUPE, a Canada-wide public-sector union. The May 18 convention called for a separate Quebec district headed by Daoust's protégé, Claude Ducharme. Reimer sent Henri (Hank) Gauthier to the convention, hoping Gauthier, a Franco-Manitoban and loyalist, would be able to repair relations, renew basic union services, and recover bargaining ground lost as a result of the Quebec section's focus on political-constitutional issues. Gauthier walked into the conference unannounced, just as a leading speaker was in full swing of a call to leave OCAW for another union. Gauthier offered his services to the locals and said he was there to help, not impose himself. He says, "Years later, I was told that they figured if I had the guts to meet them head on, they'd trust me to take on the companies."

A follow-up Quebec conference in June 1968 was primed for disaster. Ed Waddell psyched Reimer up for the worst. Some comments at the May meeting, Waddell wrote, had "only been paralleled by Hitler's racist speeches." Reimer heard through the rumour vine that Quebeckers were warned to keep up their guard, that "Reimer can sell a fridge in the Arctic circle." Reimer warmed up the meeting by announcing he'd heard the rumour and regretted it wasn't true. An igloo is warm compared to the reception in this hall, he said icily. When workers complained about understaffing, Reimer held up Daoust's time-sheets, which showed Daoust had been working for CUPE while on the OCAW payroll. You'll get a Quebec director when you establish some forceful programs and build up your membership, Reimer told them.

Reimer tried to re-establish bedrock unity by stressing common bargaining objectives. "We recognize that there is a distinct culture" in Quebec, he told the union's Canadian Council in 1969, but the union doesn't exist "for the purpose of enhancing the various points of view with respect to Confederation. Our prime motive is to deal with the employer and to establish a solidarity of working people." Reimer thought that radical nationalist politics were confined to one group among the Montreal leadership, that they didn't play in Shawinigan or even Montreal's east-end refineries. He also hoped to stave off the creation of two separate Canadian districts within the International. As long as Canadians were in an International, he thought, they needed to be a unified district. Otherwise, they'd lose their bargaining power, be open to divisions that reduced the influence of both groups, English and French. A Quebec district on its own would not carry the votes to press its needs within an International. His 1969 views are a case study of how options for relationships among Canadians are governed and limited by Canadian relations with outside powers.

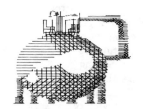

Gauthier moved to Montreal on July 1, 1968. As he started to unpack, he got an emergency phone call from BP managers, who told him he'd better settle a wildcat strike that just broke out or he'd face court charges for damages. Gauthier contacted BP union leader John More. Where's the plant? Gauthier asked. You get settled in, I'll look after it, More said. That was Gauthier's last worry-free day as senior staff rep for Quebec.

As a kid in St. Boniface, Manitoba, Gauthier had joined his teachers and classmates in defying Manitoba's English-only school regulations by smuggling French textbooks into their room. He says, "We lived in fear of the English school inspectors." In the 1950s and 1960s he demonstrated for French-language rights and worked with George Forest, the francophone leader who refused to pay a traffic ticket issued in English and got the courts to declare Manitoba bilingual.

In Manitoba, Gauthier says, "Our fights were political. In Quebec, the fight was on the shop floor." Because the bosses spoke and bargained in English, "There was no place to vent resentments and frustrations. So workers vented on the picket line." Save for the sense of distance created by his Manitoba roots, he says, "I know today I'd be behind bars. I would have been fighting with my full emotions. It was ripe. There were so many issues to the fore at the same time."

Ironically, he says, his focus on clear-cut union bargaining issues only highlighted the Quebec fact in labour relations. Gauthier went to the Great Lakes Carbon plant in Berthierville to defend a man fired for disobeying his training instructions. The worker's training had been in English. Gauthier went over the heads of local management and convinced company headquarters to rehire the man and retrain all workers in their mother tongue. While touring the plant, where the carbon hung thick and dirty in the air, Gauthier, a former lab technician, noticed that the factory's fresh air pump hung beside the exit for the carbon soot, guaranteeing a double dose of poisonous soot.

Before he could convince the local to confront this health hazard, Gauthier had to overcome the false pride that masks acceptance of

THE WE GENERATION

In the 1970s, English North America had its "me generation." Quebec had its "we generation." "We" -- or "*nous*" — was the most popular poster of the 1972 general strike, the biggest in the continent's history. Strong feelings about "us and them" fuelled the highest levels of labour militancy in the country. Clear thinking about "us and them" peaked in the most radical labour manifestos. If the climax of the decade was the 1980 referendum on independence, the climax of mid-decade was the series of referendums on U.S. unions, seen as agents for foreign domination.

Unions made fast time in the 1960s, but didn't have much to show for it at the end of the decade. "The real legacy of the Quiet Revolution was higher taxes" — up 95 per cent — Michael Smith writes in *Work in the Canadian Context*. They were mainly taxes paid for grants to help industry. But high-wage industries

still preferred Ontario. Quebec had double Ontario's unemployment rate. And French Canadians were still in lower-paying jobs, the Gendron commission showed.

We've had our fill of "promises of jobs and prosperity, when we will always be the faithful servants and bootlickers of the big shots" and U.S. millionaires, the FLQ said in 1970, after

kidnapping Pierre Laporte and James Cross. "Not for long can one hold in misery and scorn, a people once awakened." Two years later, provincial leaders of Quebec unions issued angry manifestos. Their nationalism hit at U.S. "imperialism," not at Ottawa. Their solidarity was based on socialism, not separatism. Quebec is a "service state" beholden to Bay

174

Its Up to Us

Street and Wall Street, the Confederation of National Trade Unions (CNTU) declared in *It's Up To Us*. "The workers must reclaim control of their union movement," the CNTU said. "There's more to social change than rhetoric and verbal nationalism. Workers must concentrate on the struggle against capitalism."

U.S. monopolies dictate an economy based on resources, cheap labour, and consumerism, the Quebec Federation of Labour (QFL) said in *The State Is Our Exploiter*. The monopolies "have a vital interest in destroying cultural differences between nations and in creating a uniform lifestyle and pattern of consumption around the world." The QFL called on workers, the unemployed, and students to build a socialist society. Bargaining for a bigger slice of the pie hasn't worked, the Quebec Teachers' Union (CEQ) said in *Phase One*. "It is an inhuman paradox of modern capitalist society that we speak of preparing manpower for the labour market and never of creating jobs which are worthy of men."

In 1972 the three union centrals formed a "common front behind the lowest paid workers. When Premier Robert Bourassa arrested the leaders of the three unions, 350,000 workers left their jobs. In 17 centres, including Baie Comeau, strikers barricaded roads and seized radio stations to make labour broadcasts. All through the decade, as Michael Smith's figures show, Quebec led the country in number of strikes.

Economic militancy, radical social protest, anti-U.S. nationalism —this was "the French fact" in Canadian labour relations in the 1970s second-class status. He says, "In fact, the tougher the job, the dirtier the job, the harder the job, the guys actually felt good that they could handle it." By 1973 Gauthier had negotiated to have the right to form health and safety committees included in all OCAW agreements. Later he served on the QFL committee that negotiated with the Parti Québècois on the terms of the 1978 health and safety act, considered one of the most progressive in the country.

Gauthier's focus on specific union services, as distinct from general political claims, brought tensions within the International to the surface. Although the Canadian district provided bilingual bargaining bulletins and collective agreements after 1971, the Denver office was slow to provide basic brochures and services in French. Gauthier and Reimer badgered the International to provide complete bilingual service, even at Denver headquarters. "There is a very strong new mood in the Province of Quebec," Reimer wrote headquarters in 1972. "Locals expect that people who do business in the province of Quebec must have the ability to do so in the French language."

Despite chronic understaffing and problems with the U.S. headquarters, Gauthier rebuilt OCAW as a 2,500-strong force in the Quebec petroleum and chemical industry. When the rebellion broke out against U.S.-based union structures and practices in the industry, it came not from OCAW but from the ICW.

In 1974, when the QFL won the right to charter locals on its own, independent of the CLC or individual unions, it opened the floodgates for Quebec defiance of international unions. Until then workers were reluctant to break away from international unions, no matter what their level of grievance. In industries such as construction, a recognized union card was demanded of all workers. Union activists who valued community support, political clout, and the full range of services and connections that come from central labour bodies were discouraged from leaving an international as long as it held the only pass-key for participating in central labour bodies. Once the QFL broke that monopoly, workers were free to choose their relations with International unions on their own merits. That left very little for some internationals to rely on.

In 1975 a Montreal conference of the Canadian district of the International Chemical Workers issued a call for merger with OCAW in Canada. The Akron-based president of the ICW fired all staff who acted on that conference call. The fired leaders set up a new union, the Canadian Chemical Workers, to rally workers to their nationalist cause — the right of Canadian workers to decide their own relations with other unions. The mutiny was denounced by labour leaders across the country. The former ICW leaders were expelled from all central labour bodies.

Apart from Reimer, QFL president Louis Laberge was the only major labour figure to back the rebels. In October 1975 Laberge fired off a telegram to the ICW's international president. "Your condemnable attitude and your most reproachful action will not be accepted by us in Quebec," he said. "We have been for so many years fighting bosses whose actions were more decent than yours." Despite orders to the contrary from the CLC, the QFL continued to seat Bernard Boulanger, leader of the Quebec mutiny, at executive meetings. Laberge later told delegates to a 1977 conference of the breakaway union that when the CLC formalized its demand in writing, "We read the letter as a motion to receive and file." Attacked for supporting splits and mergers, Laberge said, "This is the best accusation ever laid on me."

Laberge also joined the April 1976 press conference to launch raids against the ICW's 21 locals. Boulanger had already been hired as business agent for 15 of them. "I believe that a redefinition of the links that unite unionized workers in Quebec and Canada with U.S. workers is urgent and inevitable," Laberge told the *Montreal Star*.

In 1978 Laberge played union matchmaker. He arranged to be too busy to attend the Canadian Chemical Workers convention and sent François Gagné in his place. Gagné had split from the United Textile Workers in 1975 but continued to service his old locals under the protective wing of a QFL charter. He and the Canadian Chemical Workers, for the most part fierce Canadian nationalists, quickly found they were cut from the same cloth. "Of course, in Quebec we have to face more outside organizations compared to any other place in Canada," he told CCW delegates.

A machinist born in Hull in 1924, François Gagné descended from one of Quebec's first families. A plaque at one of the province's premier churches, Ste. Anne de Beaupré near Quebec City, notes that the site was first cleared and farmed in 1653 by the Gagné family. Another plaque near the Jacques Cartier bridge in Montreal honours Gagné's ancestors who were hung for taking part in Quebec's 1837 rebellion. Gagné inherited that rebel blood and was active in a number of union locals in the Hull area until 1958, when Roger Provost of the United Textile Workers spotted his organizing talents and put him on staff.

Textiles and clothing wear were the classic industries of old Quebec. Quebec didn't have coal, so it missed out on the age of iron and steel in the late nineteenth century. Lacking coal's firepower, Quebec attracted light industries that paid subsistence wages. Later, in the mid-twentieth century, the labour regime reflected the interests of that industrial sector. Duplessis was prepared to use police repression against unions that defied him. He was also prepared to co-operate with compliant unions that did his bidding. The United Textile Workers was such a union. It fired its two most militant staffers, Madeleine Parent and Kent Rowley, in 1952, following a prolonged strike at the giant Dominion Textiles works. Transferred to the Montreal area in 1960, Gagné was amazed by the timidity of the Dominion Textiles union local. He says, "I was told that's the way it had to be, because that's the way the company wanted things to go." Parent and Rowley had been replaced by company personnel officers. The new staffers "were good servants of people in power in the company and union, and at times there wasn't much difference," Gagné says. "The real director of the union in Canada was the president of Dominion Textiles."

Gagné was the loose thread in the union. In 1962 he was elected leader of the Quebec region of the United Textile Workers of America. At the 1965 Canadian district conference he called for unity and merger of all Quebec clothing and textile unions. "We could've had a hundred thousand members, but the problem was there'd only be one president," he says. His proposal was trounced.

Elected as one of five Canadian vice-presidents for the International union in 1972, he couldn't stomach the colonial attitudes of the union establishment. "Once, the Canadian director asked me to buy an American flag to put up at the Queen Elizabeth hotel when the International president came to town," Gagné says. At International conventions Gagné stood out when he didn't join English-Canadian unionists who stood to salute the U.S. flag and anthem. He says English Canadians "accepted American ways more easily, because they never had to fight for their survival."

In 1974 Gagné defied International instructions three times. He voted for Quebec autonomy at the 1974 CLC convention. He refused to accept an agreement with the Johnson and Johnson company that was in line with U.S. settlements. "I knew we could get ten cents more," Gagné says, "so I told the International president he should come and make his case to the union meeting in English." Then, in negotiations with Dominion Textiles, the Canadian director of the union asked Gagné to serve as his translator. "They wanted me to be a Québècois de service," (which translates roughly as "Uncle Tom" or "Negro-king") he says. In the first bargaining session, Gagné translated loosely, substituting his demands for those of the union director. When the company caught on, Gagné was removed. "When I told them I'd fight their settlement, they came after my hide," Gagné says. The director offered him a job with the labour board in Ottawa. "That's great, giving a separatist a chance to work in Ottawa," Gagné said. The director read his lips. Gagné called his supporters, who seized the union's offices until they were removed by police.

Gagné was fired on April 1, 1975. That night, two representatives from the QFL came to his house and Gagné soon began a new career as a QFL rep, servicing his old locals until the ministry of labour could supervise a formal vote to sever U.S. connections. By the end of 1976 Gagné had won over two thousand former Textile members to his QFL operation. Plant closures, together with raids by the CNTU and CSD (a conservative breakaway from the CNTU), left the United Textile Workers, once ten thousand members strong, with only token remnants from its past.

Though the QFL gave him emergency shelter, Gagné knew he had to find a permanent home. It didn't cross his mind to look to other clothing workers' unions. "They're all American unions that function with the bosses and help exploit Quebeckers. Americans are enemy number one of Quebec unionists," he says. Nor did he consider joining the Quebec-based CNTU. The CNTU had a reputation for striking first, negotiating later. More important, it was extremely centralized and gave little power to individual locals.

"I didn't want to be in an American melting pot, or any melting pot," Gagné says.

Bob Stewart of the Canadian Chemical Workers approached Gagné about a merger in 1978, but Gagné wanted to keep his options open. "Let's shack up together and see how we get along," Gagné replied. Stewart paid Gagné's salary for servicing Quebec locals that had joined Stewart's breakaway from the ICW. But Gagné had his eyes on a three-way fusion that would include OCAW. He knew Stewart had the same idea. By remaining aloof, he had more power to bargain for full Quebec rights within a brand-new organization where all relations started afresh.

Gagné had discussions with Reimer on the place of Quebec in any new organization. It helped that Reimer was a farmboy from the west and an immigrant. He had a hunch of what Quebec workers were going through. They needed, Reimer says, "greater determination to work their way through all the complexities of dealing with the church, with rapid urbanization, with being a minority. They not only had a different culture they wanted to preserve, but all these other features resisting their development. They've just been exploited more than other workers in Canada."

But as an immigrant of the pre-World War II era, when assimilation not multiculturalism was the dominant ideology, Reimer was a product of the Western Canada melting pot. If they oppose bilingualism and biculturalism, Westerners are often called "rednecks" (a term of bias, reflecting class stereotypes about those who toil all day in the sun). Some of their resentment of bilingualism is founded on "Western alienation" from the power blocs of central Canada. Some of their outbursts are founded on fear, ignorance, prejudice. Some of their anger is a venting of "displaced aggression," a contorted hostility against "French being rammed down our throats." But some of it comes from a pre-1960s progressive tradition, very strong in the West, that looked to an era when prejudice would disappear as soon as education erased small group loyalties and taught the commonality and solidarity of all human beings. A generation that suffered anglo prejudices against "peasants in sheepskin coats" (Ukrainians) and wops (labourers who came to Canada WithOut Passport), that fought the Nazis, didn't identify human rights with nationalism.

Reimer, who was the butt of "Hunky" jokes and taunts in the schoolyards of his hometown, came from that tradition. Deep in his heart he believed that the elimination of group loyalties, not their entrenchment, would lead to increased tolerance. He was an unhyphenated union internationalist. That's why he was slow to embrace Canadian nationalism, and why he looked for a "real issue" to justify Canadian independence. That's why he could promote distinctive services for Quebec but was slow to embrace Quebec nationalism. To radicals of the pre-1960s generation, nationalism smacked of narrow tribalism.

"It appears that people are finding life in society so complicated that they want to build walls around themselves and escape," he wrote in *Union News* in August 1977. "There is still the opportunity to show the world that this country, with all its diverse backgrounds, French, English, German, Russian, Chinese, Japanese, every race, creed and color, almost every conceivable religion, can live together in peace, and be an example for the rest of the world to follow. That's our challenge and that's what we accept." His article rated Quebec as one of five distinguishable nations in Canada, on a par with the west and the Atlantic provinces.

Ironically, Reimer's Western nationalism helped him understand why Quebec delegates opposed having Canada in the name of the Energy and Chemical Workers Union. Indeed, the name became the most heated controversy at the 1980 founding convention in Montreal. Former members of the Canadian Chemical Workers had recruited many of their members by appealing to nationalist opposition to U.S. union domination. They printed maple leaves and "Go Canada" on most of their leaflets. They were emotionally attached to a Canadian namesake.

Quebec unionists, by contrast, had to compete with the Quebec-based CNTU. They had to appeal to workers who were separatists. Besides, many of them were separatists. Quebec delegates came close to walking out when Ed Broadbent, the strongly pro-federalist NDP leader, gave his convention address. So Quebec militants wanted a name that highlighted their jurisdiction and union connections, not their tie to Canada and federalism. Reimer sat out the debate so he could play a conciliating role. Privately, he didn't want Canada in the union's title. He thought it was a stamp of colonialism. "When you see Canada (Ltd) on a company, you know it's a branch-plant," he says. "To me, we should be casual about the name, and strong on the content."

Despite their different starting points, Reimer and Gagné had no trouble negotiating a union constitution that gave Quebec special status. In the Energy and Chemical Workers, Quebec locals participate in all national bargaining programs, just like locals in the rest of the country. The national office provides full bilingual services. But a rank and file Quebec council, alone in the country, gets veto power over the selection of the Quebec director and union staff. "Sovereignty association is how I look at it," Gagné says.

It has worked. Membership in Quebec stands at ten thousand, despite major losses during the Textile and Chemical breakaways of the mid-1970s, and another set of losses during the severe Quebec recession of the early 1980s. Gagné, known in some circles as a "master-raider," worked with two groups that broke away from U.S. unions and joined ECW.

Cement, Lime and Gypsum workers from Quebec were the first to experiment with ECW's unique "welcoming program" for workers from other organizations. Because ECW is founded on the notion of merging unions in similar fields, it provides for affiliation on a trial basis. That way, it offers a free trial-period, in which differences can be sorted out in practice. More important, it respects the status and pride of groups, and it highlights merger and alliance, not submersion and raid.

In 1985 fifteen hundred Goodyear rubber workers from Valleyfield and Quebec City came in through that door. They were fed up with an International that denied Canadian autonomy and were looking for a new union that promised both Quebec and local autonomy. The independent United Oil Workers local at Montreal's Petro-Canada refinery also became affiliated with — and later joined — ECW under the same provisions.

No national economic or social programs have been compromised by the ECW's recognition of Quebec's special status within the union. That says something about the Canadian potential to provide unity without uniformity, to combine merger with autonomy. Politicians couldn't do it in a luxury hideaway at Meech Lake. But workers did it from picket lines and union halls.

WEST SIDE STORY

Neil Reimer was the opposite of his counterpart, Gordon McIlwaine, director of the International Chemical Workers. In 1956 McIlwaine had nixed every proposal Reimer made to get both unions working together, to follow up on the unity that had brought about the new Canadian Labour Congress. So Reimer tried reverse psychology. He knew that both unions were trying to organize a fertilizer plant in Medicine Hat, Alberta, and a chemical plant in Brockville, Ontario. Reimer wanted a clear field in Brockville and knew that ICW was way ahead in Medicine Hat. So he asked McIwaine about a swap: OCAW would get Medicine Hat and ICW would get Brockville. Reimer was sure McIwaine would refuse. But McIlwaine wasn't himself that day. He agreed. Reimer consoled himself. "At least he did the opposite of what I thought he'd do." He consoled himself a lot more as time went on. The main in-plant organizer at Brockville got fired and the drive there died. The ICW turned over enough cards in Medicine Hat for OCAW to win the local, and ICW was once again isolated from the West.

Two more contrary organizations than the ICW and OCAW would be hard to find. They were in each other's hair and at each other's throats for 20 years as their internationals tried to arrange a marriage of convenience based on their common needs and interests in the chemical industry. Then they went on a wild fling and eloped. In 1980 they merged to form the Energy and Chemical Workers, an independent Canadian union.

In hindsight the marriage seems like a triumph of logic — two unions in the same field deciding to let bygones be bygones and pool their resources. But it was a triumph of reverse logic. It defied the logic of international unionism, of the Canadian labour establishment, of the Canadian industrial relations system, and of the cultures that give unions meaning to their members. The merger took more than logic. It took a leap of faith. It took leaders who carved freedom out of necessity, did what was needed, not what was easy, comfortable, proper, acceptable, or logical. And it took a war of independence.

After Bob Stewart, a former leader of ICW, and Neil Reimer, a former leader of OCAW, resigned from their top posts with the Energy and Chemical Workers Union, they talked about the old days when they'd been at war with each other.

Everyone talks about union mergers, but few do anything about them, they said, even though merged unions mean more members, more dues, more power, more political clout, less competition in organizing, less duplication of basic services. Yet mergers are rare in the extreme. There must be more to this issue than meets the eye, they thought, some deep-seated unconscious reasons why organizations that are in the business of making more money for their members, that spout off about labour unity, don't get it together.

The first poison pill they identified was ambition. Small unions don't necessarily make for small egos. Some leaders are quite happy being big frogs in little ponds, and some prefer to submerge, not merge with, other unions. Once organizations have been formed, they're set in bureaucracy, which has vested interests harder than stone. Stewart said he used to think that union mergers were just a matter of adding up two memberships in one jurisdiction. But unions aren't faceless organizations, and they don't add up like numbers. Each union has its own body chemistry, culture, points of distinction that are larger than life. Culture was the hardest pill to swallow in any merger.

The ICW emphasized service to its locals. The headquarters had a 24-hour hotline for local leaders who needed advice. Staff reps handled bargaining, grievances, and arbitration. A staff lawyer was there for the tricky cases. That way locals had the benefit of the staff's wide-ranging experience, and kept up with the professionals on the other side. However, time spent on servicing locals was time taken away from the broader labour movement. The ICW, raised in the tradition of job-centred rather than social unionism of the craft-based American Federation of Labor, didn't mind the trade-off.

The OCAW emphasized local self-reliance. There were too many small units for staff reps to keep on top of, so staff taught locals to fend for themselves. Moreover, OCAW avoided arbitration whenever possible, believing that lawyers only got in the way of settling differences. Time saved on servicing was devoted to organizing and political action, common traits of unions with roots in the industrial union movement.

ICW was headquarters-centred, staff-based. It saw the country from its base in Central Canada, where its members worked. It bargained one local at a time. It admired the Autoworkers' approach to bargaining, which emphasized the cost of living. OCAW had a modest headquarters, far from Central Canada, and took pride in its executive board of rank and file workers. Its leaders had a Western outlook. It practised national bargaining and emphasized productivity as the basis of wage hikes.

But there was more to their long-standing friction than different cultures. The very structure of the industrial relations system in Canada kept them apart, even as shared employers in a shared industry brought them together. That's because employers go up and down, while unions go sideways. Corporate conglomerates are organized vertically, integrated from bottom to top, from the oil well to the gas pump, from the Pillsbury bun to Burger King. That's how companies control the factors of production and distribution. But the labour market is organized horizontally. That's most easily seen with skilled workers who belong to the same craft union, no matter which company they work for.

There is a great divide that cuts right through the labour market and influences industrial unions committed to organizing up and down the hierarchy. On one side is the blue-chip labour market, dominated by large capital-intensive companies that are free from cut-throat competition and can afford decent wages, even if there's no union. On the other side is the penny-ante market, dominated by small, labour-intensive upstarts that live by keeping wages lower than the competition. They can't pay decent wages even if there is a union.

The pecking order is everywhere. The petrochemical industry pays well; the plastic products industry doesn't. The logging industry sets high standards; the furniture industry can't. Central government agencies are decent employers; small social agencies pay charity wages.

It's hard for unions to cross the great divide because fear and turnover are high at the low end. Unions are hard put to swallow too much of a "logical" jurisdiction because it means indigestion from combining wage leaders and wage laggards, locals that look after themselves and make the union look good, and locals that need constant babysitting and make the union look bad.

Unions tend to seek expansion in the upscale labour market, even if that means crossing jurisdictional lines. Some unions are outright scavengers. The Teamsters and Warehousemen, for instance, say they have jurisdiction in any business that has wheels or walls. To their credit, they're among the few unions to organize across the bottom spectrum, where labour relations sometimes require a heavy hand.

The Canadian Labour Congress is a logical place for unions to find common ground around clear landmarks. The CLC's bind is reflected in its constitution. It is expected to "encourage the elimination of conflicting and duplicating organizations and jursidictions through the process of agreement, merger, or other means." But, the same article says, it is also expected to stop "raids," efforts by one union to recruit members of other unions. The two obligations work at cross purposes. That's most obvious in the transportation industry, where jurisdictional lines from the horse and buggy age have been frozen in time by the CLC's anti-raiding agreement. Likewise, dated jurisdictions in the petrochemical and chemical industries ignored the crossovers common in an era of synthetics. What corporate conglomerates wrought together, old-fashioned jurisdictions rent asunder.

OCAW tried to involve the CLC in settling jurisdictional problems. In 1967 Ron Duncan asked the CLC's committee on constitution and structure to draw up a blueprint that maximized labour unity against integrated multinational corporations. The OCAW proposal favoured a small number of broad jurisdictions covering entire sectors of transportation, construction, government, mining, heavy manufacturing, petrochemical, synthetics, and so on. That was a waste of time. So was OCAW's attempt to establish a petrochemical council housed within the CLC. Duncan asked the council in 1964 to establish a "proper atmosphere" for co-operation and merger. Reimer tried to revive the council in 1971, but to no avail. So individual unions slugged it out in the jurisdictional trenches to make sure they lived to merge another day.

These were the powerful, often irrational, forces that kept the ICW and OCAW apart and embattled for decades. It took more powerful, often more passionate, forces to bring them together. To credit the merger to logic is to understate both the problem and achievement, and miss the parallel with another Canadian shotgun merger over one hundred years earlier. People in the various provinces of British North America barely knew one another. If they did, they fought all the time. But the home country, Britain, was unable to solve their problems for them anymore, even though the United States was throwing its weight around, laying claim to the whole continent. Faced with a stark choice between continued separation and deadlock, or survival through merger and expansion, five provinces agreed to merge. In 1867, Confederation was the result.

The line between self-destruction and self-preservation was hard to spot in relations among unions with a stake in the energy and chemical field. Reimer's efforts to build a unified jurisdiction landed him in gang-wars with most major unions in the country.

In 1956 the International Brotherhood of Electrical Workers (IBEW) bested him in a contest to represent Manitoba power workers who had been members of the historic One Big Union. When the OBU joined the CLC, its legendary leader from Winnipeg General Strike days, Bob Russell, hoped the local would transfer to OCAW. Reimer counted on that endorsement to carry the membership. The IBEW attacked the OBU rhetoric and nostalgia and promised hard bargaining and a craft card that could be used across the continent. The OCAW lost the vote. "It was the first major setback in our dream to have a national energy union," Reimer says.

In the same year the OCAW lost its bid to represent uranium refinery workers in Port Hope, Ontario. Reimer told the Steelworkers' Larry Sefton to butt out of the organizing campaign, that the plant belonged to a union with "atomic" in its name. The "Atomic" in OCAW "could encompass practically anything," Sefton shrugged. "I hope your jurisdiction is not as ill-defined and troublesome to all other organizations as the jurisdiction of District 50 was," he added. Port Hope was a uranium refinery and Steel had jurisdiction for metal refineries, Sefton said. He got the plant.

In the 1960s, IBEW, with the blessing of Saskatchewan premier Ross Thatcher, snatched away linesmen from OCAW at Sask Power. When Duncan was director, he got into a series of hopeless dogfights with the Canadian Union of Public Employees and the Saskatchewan Government Employees Union over other parts of the workforce. In the 1970s Reimer got stopped by Ontario's labour board when he tried to recruit nuclear power plant workers from CUPE's operation at Ontario Hydro.

OCAW went flat out in organizing, increased in size every year, but couldn't keep hold of its jurisdiction. It didn't have the resources to carve out energy or chemical units from the giant unions. The giant unions had the resources to carve out pieces of the energy and chemical complex that appealed to them. That became more likely in 1971, when District 50 of the Mineworkers, Silbey Barrett's old catch-all union from the CIO days, gave up the ghost and sold its rights to the Steelworkers. Steel gained a major point of entry to the chemical industry and was poised for expansion. "They were going to give us the bear hug treatment, just overwhelm us with their weight," Reimer says.

What bones weren't crushed by Steel might be broken by smaller union competitors. The militant Quebec nationalist CNTU was on the rise in the early 1970s, and raided both OCAW and ICW locals. The nationalist Canadian Oil Workers Union, possibly with backing from the corporations, chipped away at OCAW's stronghold in oil.

Neither ICW nor OCAW could hold their forts any longer. It was a classic case of "defensive expansion." If they didn't unite and stake out an entire jurisdiction, they'd be squeezed out. "A merger has gone on by employers" in the oil and chemical industries "until today you can hardly distinguish one from the other," Alex McAuslane told the Canadian Council in 1963. "They merged to protect their own interests, to protect their profit position, to create monopolies and milk the public,

187

to control government at all levels," he said. "If we are going to survive," he said, "then we must learn this lesson from them."

Mergers were also "a life and death question" of fending off big unions pushing their way into the industry. "The Steelworkers are crowding us over the top in our jurisdiction," he wrote the International president that year. "Above all else, we need merger," he wrote, to "give us strength to offset the violations of our jursidiction."

By the time merger became a do-or-die question for the two Canadian unions, it was clear they'd have to work it out on their own. The track record of their internationals was hopeless.

International talks in 1963 broke down at the outset when OCAW rejected ICW demands to dump the rank and file executive board. In response OCAW's Canadian Council decided, "Failure of our International officers to act on this matter leaves us no choice but to take all steps necessary." International talks went further in 1964, but blew up at the last moment when the ICW president insisted on guarantees that he could lead the merged union for at least two terms of office. OCAW wasn't prepared to cancel elections to satisfy that ambition.

We may have to merge on our own in Canada, OCAW Canadian director Ron Duncan wrote staff in a confidential memo. But it shouldn't be "a revolution against our International Union or our American brothers as a whole," he said. "I would suggest that this not be allowed to become an inflammatory subject."

In October Duncan wrote International president Jack Knight to suggest that OCAW "should be one of the first unions to establish what would be the first truly International Union" where Canadian and all other national sections would be autonomous. "Once having made this step, we could then approach other organizations in Canada with like jurisdictions, with a view to merger." That suggestion led nowhere and at the end of the year Duncan wrote staff confidentially again to pressure the CLC to bring petrochemical unions together.

In 1966 Duncan told the new president Al Grospiron that Canadian leaders of chemical, glass, textile, and rubber unions all wanted merger and all blamed their U.S. leaders for the holdup. "I do not mention this just for the purpose of rubbing you a little," Duncan wrote, "but I think it is necessary that you know this attitude is much more open than ever before experienced; perhaps an indication of what the future trend might be."

In 1973 another International effort to merge OCAW and ICW flopped. Reimer took the bad news to OCAW's Canadian Council and said a merger would have to be worked out in Canada, "whatever the means may be." The Council set up a rank and file merger committee and told it to hussle up business.

In the 1970s the International Chemical Workers ended its aimless drift and went into free fall. When the war against Vietnam wound down, the chemical warfare industry wound down too, and membership slid from 110,000 to 80,000. The union's palatial headquarters in Akron, Ohio, was a financial sinkhole that was swallowing all the union's resources. At the 1972 international convention, dissident Frank Martino villified the entire leadership for "wilful disregard of basic principles of trade unionism" and hopeless incompetence. He won the secretary-treasury and got appointed to the top job in mid-term when the president was forced out for alcohol abuse.

In Canada, the local leaders of fifteen thousand chemical workers demanded decisive action. Their wages were barely keeping pace with average industrial wages, despite high productivity and profits in the chemical sector. They knew they were stuck in a rut, as both a bargaining and a social force. Delegates to a 1973 national conference broke into stormy applause when a CLC official spoke in favour of merger with OCAW. "It appears to me that we are going to have to make a crucial decision with respect to our relationship with ICW," Reimer wrote OCAW president Grospiron. "Some of the membership of ICW will go any direction if it becomes clear to them that their Union has no intention of working out a merger."

The ICW's Canadian vice-president, Thomas Sloan, was a "super guy," but he did "nothing to give some sort of direction to the organization," said Bob Stewart, who had become Sloan's assistant in 1968. Every morning Sloan broke into a cold sweat waiting for his phone call from central headquarters. Then he'd relax over a coffee and attend to his orders for the day. He followed the course of least resistance with his 92 small and dispersed locals. The union didn't launch company-wide bargaining programs until 1968. Sloan carried out international policies faithfully but "there wasn't anything tailored to the Canadian situation," Stewart says. "It was just everybody for himself."

Sloan's style reflected the union's American Federation of Labor heritage. AFL unions bargained on their own and struck on their own, often against the advice of head-office staff who worried more about the drain on strike funds than bargaining precedents. "Most times, what happened was we'd get all heated up," says Welland Cyanamid leader Al Wood, "and then a telegram would come in from Akron saying, `You've got to cut this thing off. It's becoming too real.'"

Members had few chances to set union policy. At ICW conventions staff members walked in carrying fistfuls of proxy votes on behalf of locals that couldn't afford to send delegates. Politics centred on squeeze plays for union jobs, not policy directions. Sloan was edgy about his job, lived in fear of his president, because he knew how cut-throat the game could be, that his career depended on the central leadership, not the membership.

Bob Hammerton, another leader from Welland Cyanamid, was in a convention hospitality suite when a staff member waved him over to the bar, out of sight of a delegate relaxing in an easy chair. What do you think of Gord McIlwaine, the Canadian leader? the staff member asked Hammerton. "He's got no guts, does whatever the international tells him to," Hammerton said. When Hammerton finished cutting up McIlwaine, the rep asked him to sit down — "And," Hammerton says, "who the hell was sitting in the damn chair but McIlwaine, tears running down his face. I looked at that damn rep and I knew that I'd been conned." The convention sent McIlwaine into retirement and selected Sloan as Canadian leader.

By 1974 Sloan was too sick and tired to buck the system. He was also too nervous about being shown up by any young Turks on his staff. Reimer stepped up the pressure for merger when Sloan delivered his routine fraternal address to the January 1975 meeting of OCAW's Canadian Council. As long as we're divided, we can't keep up with organizing an industry that's continually expanding and moving, Reimer said in his reply to Sloan's speech. "Talking merger ten years down the road is bullshit," Reimer said.

Sloan agreed. "The crux of the thing is the potential to organize and if we can't or don't do it, the poachers will," he said. Then Reimer popped the big question. "If our International Union doesn't move off their butts, then we should merge ourselves here in Canada," he said.

Sloan backed off. "I can't see how we can form our own unions here in Canada. We would have a lot of difficulty," he said. He promised to raise the matter with his union's international executive.

Sloan missed his chance. His staff couldn't take his indecision, his submission to the International, any longer. Stewart started organizing members, getting them to press for action. He was accused of making allegations against the leadership. No, just revelations, he countered.

Stewart was the union's senior statesman, on staff since 1953, assistant director for almost a decade. He had energy to spare. Moreover, he and President Frank Martino were tight. Stewart had nominated Martino for secretary treasurer in 1972 and swung the Canadian vote to put him over the top. After that, Stewart says, "We'd go out together, we'd go to the races together, we'd go out to dinner together, we'd go to the bar together, and we were just real friends."

But Stewart was his own man. He had a deep sense of labour's role in the community and was active on the school and hospital board in his hometown, London. When scratched, there was a repressed lawyer under his skin, but deep down he liked to think philosophically, to fit himself into a bigger picture. In the 1970s he took time out from union work to study law and think through his basic goals. "And then I had to take a look at me, as a trade union leader, and what my obligations were," he says. It made him look upon merger in a new light, not just as a rehash of empire-building rhetoric.

In Quebec, staff rep Bernard Boulanger looked to Stewart to lead a palace revolt for Canadian autonomy. Boulanger was an important leader of the Quebec Federation of Labour, highly regarded for his opposition to Trudeau's War Measures Act and for his militant approach to health and safety. But he wasn't a prophet in his own International. He got called on the carpet at Akron for his "Communist" activities. It was Stewart who stuck by him, not Sloan. Boulanger thought, "It was time we got our day in court."

When ICW staff met in April 1975 just before their Montreal conference, the complaints were so heated that Sloan retired on the spot and left. Stewart took over as director the next morning, and without ado set the agenda on merger. When a man has promised to marry his girlfriend for 20 years but keeps putting it off, she starts to worry, he said. It was time to take the leap. The rest of the conference ran itself. The only debate was over whom to merge with. Delegates looked to a combined union made up of the ICW, OCAW, and the Rubber Workers.

Stewart called Reimer and suggested a get-together. At the very least, the staff of the organizations should meet and find out that the other ones didn't have horns. Gasworkers in both unions pressed hard for their staffs to work closely in bargaining. Reimer invited Stewart to speak to an OCAW council meeting in September 1975.

Stewart was used to giving speeches. But on September 16, the day of his talk to OCAW, he was nervous. He had a prepared text that took two minutes to read. His Canadian executive had approved it, though they told him he'd be fired for sure. He paced the floor of his hotel room, read it to anyone who'd listen, read it to himself in front of the mirror. Then he went to meet the OCAW delegates. Workers can't remain divided when monopolies are interlocked and centralized, he said. "The game of resolution passing and eloquent speechmaking is over. The time for action has arrived."

Then he laid it on the line. "To us, a merged national union is more attractive than an unmerged international union. The choice will be theirs," he said. Merger won't happen "unless we take some coercive action to move it along."

Stewart hadn't cleared his speech with Frank Martino and Martino learned about it the wrong way, when an OCAW officer saw him at a meeting and asked for his reaction. To Martino, Stewart's approach was blackmail. Martino started working the phones. He and Stewart had been so close that he knew all of Stewart's weak points, all the sore points of his staff. He phoned Boulanger, told him to book his family into a resort, courtesy of the International, told him he was the choice for next director. "Tomorrow morning when I shave, I don't want to be ashamed of myself," Boulanger told him. Martino kept phoning. He pried loose some backtrackers.

On October 22 Stewart went to the International's regular executive board meeting in Akron. He checked in at the hotel, dropped by the bar where everyone hung out the night before meetings. No one was there. Another vice-president passed him in the hall, pretended not to see him.

Stewart went back to his room. "Here I was, a lonely guy, like a skunk at a garden party. And that hurts, you know, I'll be quite honest with you, that really hurts. I thought we had real friendships, but nothing you did counted." The cold shoulder is a powerful weapon in a movement that relies on camaraderie.

Next day at the board meeting Stewart saw that he was fifth on the agenda and began to breathe easy. They wouldn't let him stay that long if they were going to fire him. Then Martino came in and moved that Stewart be put on the top of the agenda. "What have you got to say for yourself?" Martino asked. "What good does it do to talk if you've made up your mind? Stewart asked. "I'll make up my mind fast if you don't talk," Martino snapped. "Have you made up your mind?" Stewart repeated. "Yes," said Martino, tossing Stewart a termination letter and telling him to hand in his keys to the new Canadian director, Tom Sloan, who would be brought back from retirement. No one said a word when Stewart walked out. Stewart phoned Reimer and some of his staff, then drove back to Toronto, stopping off in London to tell his wife he'd been fired. Martino forgot to change the headquarters locks, which gave Stewart's people the weekend to photocopy membership lists and make long-distance calls to locals they serviced.

Martino sent telegrams to all the Canadian staff telling them not to meet with Stewart, on pain of instant dismissal. Rolf Nielsen, just transferred to Calgary, got his telegram and jumped on the overnight "cardiac special" to meet Stewart in the morning.

Martino flew to Toronto to meet all staff. Neilsen says, "He wore his famous blue suit with a white belt and white shoes, and poured some drinks." Martino told the staff they could "bank on Frank." The staff said "Thanks for the drink, and we'll see you." Telegrams firing them were sent to their homes that night. Because they'd been late renewing their staff collective agreement with Akron — a regular and usually inconsequential event in Canadian labour relations, but grounds for terminating bargaining rights in the United States — Martino denied them access to any grievance procedure.

The ferocity and pace of events took everyone aback. Those who stood by Stewart's stance on merger were running on nerve, and they started to falter. Their lawyer, Ian Scott — later, Attorney General of Ontario — told them all but one breakaway from U.S. unions had failed. He advised them to negotiate a settlement. He called Martino, who agreed to rehire Stewart as a junior. The proposal went the rounds of the rumour mill at the Ontario Federation of Labour convention. Everyone, including nationalist leaders of independent Canadian unions, said Stewart and his allies were crazy not to take the deal. Stewart put a stop to the debate. "This isn't about me. This is about merger," he said. "This was all done to say you either do it on an international basis or we'll do it in Canada, and let's not lose sight of that — that's what it's all about."

The finality of what happened dawned slowly on Martino loyalists too. Reimer met Sloan in Alberta on November 3, 1975. Sloan expected to lose three thousand people at most, to come out the better for it, to take no losses in Alberta. That night, locals that Rolf Nielsen had serviced made their plans to keep Nielsen on retainer and wait for the right time to bolt. No one had foreseen the turn of events. Stewart hadn't when he took out a loan for a boat and pool that summer. Nielsen hadn't when he took a transfer to Calgary and his wife had quit her job.

Reimer had long believed in carrying out mergers on the basis of broad principles and ironing out the troublesome details once the pieces were in place. He hadn't expected a detail like the fact that Stewart had no members to merge. "The firings created a problem insofar as merger was concerned," Stewart says. "Instead of working from within, we were faced with bringing about merger from the outside." There were no structures, no rules, no guarantees, just a battle of wills and a leap of blind faith.

Within weeks Reimer got ahold of $50,000 from Grospiron and $50,000 from his Canadian Council to help bankroll Stewart's new organization, the Canadian Chemical Workers. He gave the money away, no strings attached. For Stewart and his group, the emotional buildup of years of competition with OCAW and their fresh hatred for all international unions made a direct jump into another international impossible. On November 19, both groups signed a "Chelsea Manifesto," as in Chelsea Hotel. They agreed to merge in the indefinite future in a new organization, as yet unnamed and with a constitution and international affiliation as yet undecided. Stewart's group needed space, and it needed time to win ICW locals to a breakaway union, the Canadian Chemical Workers.

The ICW and much of the Canadian labour movement were ripped apart. On December 14 Martino called a meeting of ICW locals in Toronto. Don Montgomery, secretary treasurer of the CLC, and Lynn Williams, national director of the Steelworkers, offered support for Martino from mainstream labour, but couldn't be heard over the heckling and jostling of Stewart supporters. Except for Louis Laberge in Quebec, official labour backed the ICW. Both the CLC and the Steelworkers loaned staffers to Martino to tide him over while he hired replacements for the people he had fired. In April 1976 Montgomery sent out a memo expressing the CLC executive's "full support" for the ICW and the expulsion of Stewart's "splinter group" that "left" the ICW. Montgomery spoke to the ICW convention later that year and Martino thanked him: "We would certainly love to make you an American citizen." Montgomery said his mother was American. "There you have it. That proves why he was okay," Martino told the delegates.

No one in the CLC criticized Martino for violating the CLC's 1971 and 1974 guidelines on Canadian autonomy. Those regulations should have protected Stewart's exercize of a director's decision-making role and should have prevented Martino from firing or appointing a Canadian director. The CLC's failure to stand by its policies reflected the pressure of the building trades unions — which openly defied the guidelines — and of powerful industrial unions wed to their internationals, the Steelworkers in particular.

As a former Steel staffer, Montgomery may have sensed the union's feelings on the subject. The *Financial Post* speculated on May 22, 1976, that Steel backed ICW "because it has a vested interest both in preserving a membership it may one day incorporate, and in discouraging a similar movement within its own Canadian ranks."

In the United States and Europe, steelworkers' unions usually claim jurisdiction in the chemical sector. Throughout 1976 the international presidents of the Steel and Chemical workers were engaged in merger talks. So Williams got very indignant when Reimer asked the CLC's permission to raid an ICW gasworkers' local that worked side by side with an OCAW local in the same company. Reimer's request was "such a hypocritical act and so inconsistent with anything that one would use as trade union morality or rationality that I could not comprehend it," Williams told the ICW's *Chemical Worker*.

Whatever the CLC's position, it exercized little influence. In January 1976 the ICW went to the CLC to accuse OCAW of aiding and abetting a split and defaming the reputation of another affiliate. Reimer countered by proposing a review of obstacles to unity of the chemical jurisdiction. This would be a more comprehensive and positive approach to take, he said. It was also more time consuming. The CLC's committee took a year to report and "seriously criticized" OCAW and others who overrode union principles "because of a desire to be a winner in a numbers game." No sanctions were imposed. Though both OCAW and CCW endorsed the report, no group followed its recommendations. By the time the report was issued, ICW's fate was sealed.

While Reimer stalled, the leaders of the new chemical union survived with their backs to the wall and fire in their belly. Like the union evangelists of old, they had no money. Unlike them, they had credit cards. "Once we'd run our Visa cards to the maximum, we tried to get some other credit cards," says Kenny Rogers, who was a CCW staffer at the time. "I suppose we can thank the Royal Bank, the Bank of Montreal and so forth for financing us."

Individual locals found ways to keep their funds from being impounded by the ICW. In November 1976, Don Montgomery offered the Cyanamid local in Welland temporary refuge with the CLC, prior to a transfer to the Steelworkers. "Rather than wait and end up with the USWA," the local's minutes of November 22 record, president Al Wood and his treasurer took out $160,000 from the local account, leaving only $6,000 to cover a Christmas social. They took the $160,000 to another bank and put it in a numbered account. "It'll take them a minimum of seven or eight years to find the money," the manager said. Three days later, ICW put the local under trusteeship and seized its paltry assets. Wood threatened to go public with charges that ICW had stolen money set aside for kids' Christmas toys, and the ICW left empty-handed.

As soon as ICW collective agreements expired — that's the only time when workers can switch to a new union — the CCW launched raids. They hit hard with a pure and simple nationalist line. Their leaflets were headlined "Go Canadian," and highlighted a Canadian flag. "If you believe that Americans should run your union in Canada, then you should stay with the ICWU," they said. ICW is an American, not an international union — "Don't let the pro-Yankee ICWU defenders tell you any different." Leaflets made no reference to merger.

In 1976 this straight-up nationalism was worse than hitting below the belt. It was narrow, negative, strident, adolescent, reactionary, bourgeois, beside the point, and beyond the pale. This was almost a decade before the Autoworkers made independent unionism legitimate, before unions worked in coalitions against economic continentalism. It was only a few years after unions had ordered the NDP to expel the Waffle faction, which had attacked U.S. corporate and union domination of Canada.

In the United States, Martino circulated the CCW's anti-American leaflets and Grospiron was badgered by other international leaders for giving aid and comfort to the enemy. Reg Basken, president of the Alberta Federation of Labour and an executive member of the CLC as well as Reimer's assistant, was pressured to disown his ultra-nationalist allies.

But nationalism gave the CCW a fighting chance. "This was an organizing campaign and you've got to tell the people what they want to hear," says Stewart. "So we used it because it was our most effective weapon." The CCW blasted the ICW off the map in certification fights. CCW won the Gestar Chemicals plant 120 to 4, the Exelon plant 188 to 4, a Domtar plant 39 to 2, a Continental Can plant 86 to 4. At Cyanamid in Welland, one man signed up the entire plant in less than 24 hours. "All you had to do was say it was totally Canadian, they didn't care who operated it," says Al Wood. "They just wanted Canadian."

CCW staff worked at breakneck speed, "running around at 50 years of age like you're a young kid," Stewart says. "He wanted you to be on the job 25 hours a day, and he was on the job 26," says Boulanger. "How could you refuse him?" They passed out leaflets at plant gates at sunrise, serviced locals during the day, held meetings in the evening, and then handed out more leaflets to the incoming graveyard shifts. Somewhere, they found time to organize Ontario's first union local of bankworkers in Simcoe.

Stewart's band was merciless with ICW staff, treated them as beneath contempt. ICW staff members were afraid to show their faces around the plants, Nielsen says, and that "really left the field open to us to raid, or 'liberate,' as we were saying." The ICW went through three directors in two years. Many staff members spent half their time looking for other union jobs. With enemies like that, the CCW didn't need friends. Perhaps the CCW came on too strong for its own good. The ICW fell apart faster than the CCW could pick up the pieces, keep the membership intact. Unions that Reimer calls "vultures" picked away at the bones. The Paperworkers took three locals.

The biggest blow to CCW hopes came in November 1976 at the giant Toronto Kodak plant of twelve hundred workers. Kodak, notorious for its picture-perfect paternalistic labour relations and commitment to a union-free corporate chain, made union life difficult at the best of times. The union was on the ropes after a disastrous strike in 1974, when local leaders let prominent labour lawyer Jeffrey Sack pose a series of unrealistic demands around contract language that few workers understood. The Steelworkers, enjoying Martino's trust, took the local from under his nose, with promises that a big union could handle the company. Williams had apparently grown tired of the merger waiting game and decided to pick the local up before it was lost to the CCW.

Martino filed raiding charges against Steel with the CLC, but got no satisfaction. Disgusted, he pulled the ICW out of the CLC. He sought protection from the Teamsters, who offered him a mutual-aid pact in 1977. In 1979, 150 delegates representing 5,500 workers who remained with ICW voted unanimously to fall in as the chemical section of the Teamsters. Martino got the equivalent of a finder's fee, 20 cents per member per month to cover health and safety research.

Once Martino left the CLC, he fell from Montgomery's good graces. Martino was "insensitive to Canadian opinion," Montgomery told the business magazine *Chemical Week* in June 1978. "It's a case of sour grapes. Martino had handled it badly and was looking for a scapegoat, and the CLC was around." The Steelworkers cried sour grapes earlier that year, when Kodak workers opted for a company union. Steel did what it could "to put some blood in the veins of the workers," a staffer told *The Globe and Mail* on February 24. "They are serfs as far as I am concerned. They are scared to death of the company. They don't know what a union is all about." The workers had been without a contract for a year. Steel lost over $50,000 in its venture.

Martino wrote a poison-pen letter to Williams a month later, accusing Steel of playing into management hands. "You are quite familiar with the continuing resistance of millions of workers in both America and Canada to joining Unions like ours," he wrote. "From where I sit, one of the more subtle reasons for that resistance is an underlying awareness that Unions themselves are willing to adopt a so-called corporate ethic which says `Anything goes as long as we can get away with it.'" This was Martino's death-bed conversion. He had previously conducted himself like any other corporate executive who brooked no challenges from underlings.

No one came out ahead. The CCW turned the issue of Canadian autonomy from a series of abstract resolutions into a crusade for membership control of unions. But the union wasn't able to do what it set out to do — unify and strengthen workers in one jurisdiction. It was only able to salvage half the original ICW membership for a later merger that gave birth to the Energy and Chemical Workers.

Stewart still takes the hurt of his ruptured relationship with ICW personally. He had worked full-time for two decades to build the union. "After doing all this," he says, "then you leave and under circumstances that aren't very happy. You know, a success story with a tragic ending."

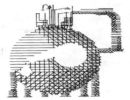

In October 1978, legal staff and leaders from OCAW's international headquarters met with their Canadian counterparts at the Bayshore Inn in Vancouver to hammer out new directions for the union. At dinner, John Tadlock, Grospiron's chief counsel, had a few too many drinks and got out of hand. His language got foul and he ridiculed the Canadians' airy-fairy talk about Canadian identity.

Reimer fell silent. "It's an old trick. If you get into a fight, you talk about what you fought about, rather than the issue," he says. But finally he had enough, and the dinner broke up.

The Canadians went back to Basken's room to polish off some whiskey and think of things they might have said. Then Grospiron knocked at the door and came in with a big grin on his face. "I fixed the sonufabitch," he said. "I got him right here," he said, pointing to his rib cage. "I flattened him on his bed. I'll teach that bugger to insult my friends."

That was the way business was done in the OCAW. They were strapping men with tough exteriors, thick skins, cast-iron stomachs, and scrappy manners. But strangely enough they also had the most unisex union style on the continent. They respected each other, listened to each other, nurtured each other, and were committed to working out understandings, even hard and sad ones. They didn't like to walk away from their differences. As a result of this culture, OCAW became the first — and by 1990 still the only — international union to volunteer independence to its Canadian district. The Canadian district became the first — and by 1990 the only — independent Canadian union to forge a truly international relationship with U.S. workers. The style took a little longer, but it took.

The Bayshore Inn meeting was supposed to tie up loose ends left over from a constitutional amendment passed by the 1977 International convention. The new clause stated that uniform administration of districts in both countries was not always feasible. There were no specifics. The wording was deliberately vague, it was permissive, it left options open. It empowered union leaders to establish a "Canadian identity," with its own structures and policies, but retaining "organic" relations with the International.

Reimer had gone to the 1977 convention knowing he had to bring home the power to set up an independent Canadian operation that could conduct a solo merger with the breakaway CCW. U.S. delegates weren't partial to that, says Rick White, Canada's representative on the constitutional committee. "They felt we were breaking up a friendly house."

Reimer soothed the U.S. delegates, trying to explain the Canadian proposal as a step towards genuine internationalism. "If our intent was to break away, we could have done that a long time ago," he said, opening the debate. Canada was a colony of U.S. multinationals, he said, and Canadian workers needed international ties to keep up with the companies. "This is not a time for restriction, for isolationism or nationalism." Without much help from the executive, the Canadians carried the day and won recognition of the Canadian union's complete independence.

Grospiron had the best of intents when he launched post-convention discussions on future relations with Canadian leaders. He owed his election in 1965 to the solid Canadian vote. Canada was his favourite district, the one he counted on to carry progressive policies. But for the life of him, he couldn't figure out what the Canadians wanted. That's why he had his lawyer draw up a questionnaire, to get hard and fast answers to 27 questions. Would this create an autonomous union? he asked. Yes, the Canadian district would be self-governing. Would there be a continuing relation with OCAW? Yes, there would be a partnership, shared research in health and safety and bargaining strategies, a joint strike fund. The two answers didn't jibe for him. A bit like the English-Canadian reaction to Quebec proposals for sovereignty-association; wasn't it just a sneaky way of saying separatism?

Leaders on both sides had it out at a Vancouver meeting in October 1978. "I've never understood two words — identity and organic," Grospiron said, passing out his questionnaire. Organic means living, Reimer tried, it's not nationalistic. "How much is this fucking thing going to cost?" a U.S. member butted in. Organic means what you have here, based on a Canadian identity, Buck Philp piped in. Philp was trying to be helpful. Unfortunately, part of the Canadian identity is that no one can define it. "I've been extra nice today," said Grospiron. "We now have an organic difference in the U.S. and Canada."

Grospiron tried again. "What you said leaves me with the impression there will be two distinct organizations tied together somehow," he said. "If you took the document before lawyers and constitutional bodies they would throw up their hands." Basken, who took minutes for the meeting, jotted down a note to himself: "It all depends upon whether or not you want to build something or find reasons for not doing it."

Grospiron was getting flustered. "You seemed to want to block something out of your mind. I'm not sure you know what you want," he said. How does this mixture of independence and sharing strengthen International OCAW in Canada? It expands the range of activities, said Reimer. Grospiron gave up. "I have no control," he said. "We would have no jurisdiction left." He reaffirmed that separation was alright by him. "If the meeting has served to convince you we want to separate, we have failed," Reimer said.

They were speaking past each other. The Canadians were talking concepts, or as they put it, "postures." The Americans were talking rules. Maybe that's a legacy of different political heritages. The British system, which has deeply influenced Canada, paid no heed to constitutions — at the time of this meeting, Canada hadn't yet been "chartered" — and relied on evolution, on adapting precedents to life. That's what organic is about. The Americans laid out their rights and purposes in a Constitution that holds for all times and places. That's led to a fixation with formalities. It's what artificial preservatives are all about.

It was "the most frustrating time of my career," says Basken. "We wanted to maintain an international union but establish clear independence. We talked about a social change, not just a border separation. We saw a new world of individual identity and international co-operation."

The next month, Reimer wrote Grospiron and made one last try at setting out a middle "internationalist" option, somewhere between a detached Canadian union and an attached North American one. Reimer wrote, "If it has been the posture of the International Union that Canadians can separate, and we would do so on friendly terms, then why in the world is it not possible for us to seek a new structure, whereby we preserve the heart and spirit of our relationship and simply have a shift of some authorities?"

Reimer gave formal answers to Grospiron's questionnaire but warned, "If this proposition fails because of technicalities, then we do not think that history will be a gentle judge." International unionism has to be based on projects that help workers reach common goals. "If we perceive of our linkage as being one of constitutional power and authorities, then that perception is wrong. It does not exist. Our linkage is the association of people."

Reimer couldn't get through to Grospiron. "It was just that Grospiron didn't want to see the break-up of the family, " says Charlie Armin, long-time friend to both men. "It's like a child leaving, and Grospiron had a very paternal feeling for the membership in Canada. So if there was one thing that I think kind of tore him up, it was that."

In the absence of a positive U.S. response, OCAW's Canadian Council was prepared to go it alone. Reimer gave his report favouring complete independence and it was seconded and passed unanimously without debate. Everything had been said, the choices had been made. For Reimer the anti-climax brought to mind the non-debates on motions to merge the Canadian Congress of Labour and the Trades and Labour Congress in 1956, ending decades of intense factional war, and the CLC's 1960 decision to back a new political party, the NDP. Unionists prefer to argue over small things, he says.

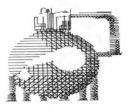

Grospiron tried to delay the final reckoning beyond the 1979 International convention. He was stepping down then, and everyone expected a hot contest to choose his successor. He told Reimer the risks were too great that the Canadian issue would become a "political football." But Reimer couldn't hold off the CCW any longer. In spring 1979 he met with the two presidential candidates, Bob Goss and Tony Mazzochi. Both agreed to support the Canadian proposal, and to stay clear of the controversy during their campaigns. In return, Reimer agreed to lift Canadian caucus discipline for the presidential vote.

In the heat of August, at the Diplomat Hotel in Hollywood, Florida, the convention sweated out a photo-finish race for the presidency between two candidates of very different views. Bob Goss was the administration favourite, a long-time staffer originally assigned to the union's international work. He was typecast as the bread and butter candidate. Tony Mazzochi, the rebels' favourite, pledged to "take the cobwebs from our soul" and rebuild the union as a community movement. He was typecast as the bread and roses candidate.

Canadians tried to put the best face forward on their apparent failure to win an organic relationship. A slick brochure, "Build a Strong International," made the case for both Canadian independence and continued sharing of the burdens and benefits of an international strike fund. "Canada is a nation searching for its own identity," the brochure explained. "It is attempting to remove itself from colonial status — culturally, economically and politically. This has profound effects on all our institutions, including the labour movement."

Why don't we just go away? "This would be the simplest thing to do," but also the truly colonial thing to do, the brochure argued. "Canadians feel that the International Union belongs to them just as much as to anyone else and they have the right to change the constitution and the method of operation. To suggest anything else represents a paternalistic and `colonial' attitude and would not be consistent with the word `international' in our name."

Delegates were won over by the Canadian logic, but remained sullen. "It's like a divorce from a wife you still love," one U.S. delegate told a radio reporter. "If she feels that's what she wants, you will very reluctantly agree and let her go to the man of her choice. You wish her the best, but that doesn't mean you give her the house, the car and a lien on your pay."

Roy Barnes from Texas voiced these emotions from the floor, but Basken countered with a powerful speech that floored his audience. Grospiron walked up to him afterwards and whispered, "I should have fired you when I had the chance."

Then Grospiron walked to the podium and called for the vote to determine whether an independent Canadian union centre could be set up. Unless the voice vote is unanimous, he explained, a roll call is required for all constitutional changes. The motion was put and Grospiron raised his gavel and asked for nay votes. Someone coughed. The motion was put again. Grospiron raised the gavel for a count of five. It felt like an hour, Basken says. "I'm sure there will be a celebration in the old home town tonight," Grospiron said. Canadian and U.S. delegates hugged each other. Many had tears streaming down their face.

After winning the right to set up their own house after the convention, the Canadians voted for the next International president. It was an open vote. Each delegate cast a number of ballots, depending on the size of the local represented. Districts voted in turn, and it was a see-saw contest. Canadians were last to vote. The count so far was already tallied. It was close, but Mazzochi was going to lose. Pressure mounted among the Canadians to split the vote, a gesture to ensure good relations with whichever president would end up negotiating the terms of severance. The entire Canadian vote went to Mazzochi, bringing his support to 49 per cent. The Canadians went out on a high note, Americans said, firm on principle to the end.

Grospiron didn't take retirement well. He moved to the wilderness of Wyoming and cut himself off from his old friends and associates. One day in 1981 he pulled his car over to the side of the road, took out a pistol, pointed it behind his ear, and fired. The shot was so clean that a passerby thought he'd had a heart attack. Doctors didn't figure out what had happened until Grospiron was put on the operating table. Grospiron, apparently, liked his goodbyes neat and final.

In April 1980, 350 delegates from what used to be OCAW, CCW, and an independent Quebec union of textile workers met in Montreal to form the Energy and Chemical Workers Union. With thirty thousand members the new union unified the major groups of energy and chemical workers, merging the experiences of the shock troops of Canadian nationalism, partisans of Quebec nationalism, and champions of classic internationalism. Bob Goss brought greetings from OCAW. Canada suffers from too much U.S. domination, he said. "There's no reason why Canadian workers can't have their own unions." Agreement was reached on plans to share research, health and safety services, and a strike fund.

Delegates from the John Mansville asbestos plant in Scarborough, Ontario, returned from the convention to join picket lines in a strike for health and safety protection, the first strike in Canada to lead to a shutdown of a killer industry. The strikers, former members of the ICW, had never paid a cent into OCAW. At the end of their first week on the line, they received strike pay from OCAW headquarters in Denver.

APPLIED CHEMISTRY

On the job

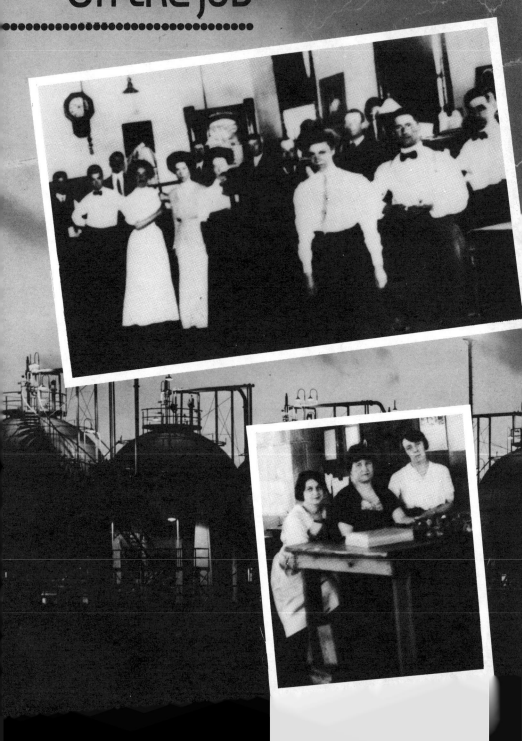

THE BEST THINGS IN LIFE ARE BARGAINED

From a labour relations point of view, the problem with life is that it's priceless. Put another way, it's free. Either way, it's hard to bargain. Unions were set up to bargain the wages of labour, not the wages of sin — death. But they've been forced to bargain health, safety, and the environment because of the failures of organized medicine and government, along with the cold-blooded way business keeps its books.

Health professions walk away from the problems. Few company doctors are trained in industrial medicine, and fewer still report to workers. Toxicologists and hygienists who set regulations on industrial poisons are the hired guns of business, according to a study in the 1988 American Journal of Industrial Medicine. Family doctors get paid to treat illness as a problem of the body, not the environment. Surgeons win glory by sewing up organs, not unravelling connections that cause organs to malfunction. Even public-health specialists stick to germs and social diseases, not poisons and economic diseases.

The accounting system used by governments and corporations hides health and environment costs. Governments think they make revenue from tobacco and liquor taxes when they're really making a killing — and paying out more for health costs. Workers think they're getting good wages and cheap prices, but don't count the hidden subsidy paid by their bodies and by workers in Third World countries. Corporations finance their profits out of unpaid debts to the environment they've pillaged. Deficit financing, not redistribution of wealth, has paid for North American prosperity. By the time the bills came due, it was too late.

Neither unions nor workers are built to spot the way the books are cooked. Workers have a vested interest in life, but they also have a conflict of interest: They need jobs. They'll bet their life to keep them. Because it's not a deal they like to think about, they repress their fears. It's called avoidance. Few union careers are built on health and environment issues. No one likes bearers of bad news. If the news is good — that injuries don't happen — no one notices. Until recently unions couldn't afford health, safety, and environment specialists. They went after money and delegated "soft" issues

▼ PART II
▼ SECTION I / ON THE JOB
▼ CHAPTER I

THE FILTHY RICH

to legislative departments of central labour bodies. Even there, health and environment had to compete with pensions, human rights, and medicare.

Health and environment are sidelined by the industrial relations system itself. Unions are tolerated as long as they know their place at the bargaining table. They can't infringe on "management rights." Reg Basken, OCAW's first health and safety co-ordinator, calls this "slot machine unionism, where workers put their dues in, pull the handle, and catch the coins."

According to Basken, "Companies don't want a union that has an ongoing role in workplace design. That's why health and safety committees were so bitterly opposed. They were afraid to give us greater respectability and involvement in the workplace."

When Reimer was a boy he helped a veterinarian castrate a stallion on his dad's farm. "This guy told me to put a tourniquet around the horse's lower lip and squeeze. While I did that, he cut off the testicles and walked away."

That left a lasting impression for Reimer. He says now, "The pain in the horse's snout was so great he couldn't feel the other pain. Just like unionists today react to the pain of low wages, while our nuts are being cut out by poisons in the workplace and environment."

The comparison is not that far-fetched. Early managers learned their manners in the stable: It says so in the dictionary. *Manage:* "(archaic) the schooling or handling of a horse or the technique of such schooling and handling," says *Webster's*. This may explain why it's second nature for managers to keep people howling about money to the neglect of more seminal issues.

The Oil, Chemical and Atomic Workers was one of the first Canadian unions to poke away at all the sore spots created by industry, to see safety, physical health, mental health, and the environment as an indivisible set of issues. That's partly due to the industries they worked in. The dangerous equipment and products workers handled forced them to be safety conscious, or risk being blown up. As chefs in the chemical soup-kitchen of the world's toxins, they were first in line to see the connection between damage to workers' health and harm to the environment. The psychological wear and tear of machine-dominated workplaces, a central feature of advanced "processing" industries, led to concern for mental health in the workplace.

Not that the union acted as fast as was needed or possible. The deadly effects of lead and asbestos have been known since ancient times. Alice in Wonderland wasn't the only one who knew about mad hatters — so named because of brain damage caused by the mercury used to make felt hats. Tar has been known to cause cancer of the testicles since a 1775 study of chimney-sweeps. Murderers and suicides have long known about the effects of gas. It took until 1969, the year the first man landed on the moon, before the OCAW took a giant step to bring modern science down to earth. When it did, it shook the foundations of Canada's industrial relations system and helped launch the modern environment movement.

The Depression generation couldn't afford to be overly careful about workplace safety. "I knew a lot of World War II veterans who'd gone over the trenches and flown fighter planes, but when they got home they were afraid to join a union," Reimer says. People would risk their lives in war but didn't seem to want to risk standing up to the boss. Dread of layoffs was tatooed on the soul of every worker raised in the Depression. It spurred them on to fight for seniority, for union security, for social security. But it scared them to death to push too hard on safety, health, or the environment.

Workers at Toronto's Consumer Gas shops won their first collective agreement in 1940, following a two-year campaign by union organizers to get governments to deal with widespread skin cancer among shop workers. The city's health department traced more than a fifth of the province's skin cancer cases to these shops. A follow-up study by Ontario's health ministry in 1941 pressured the Workmen's Compensation Board to recognize skin cancer as an industrial disease and compensate victims in 1942.

But workers, who put in standard 70-hour and 80-hour work weeks, resisted their union's fight for shorter hours. They complained that they'd make less and spend more if they only worked six days, Gil Levine notes in his 1948 M.A. thesis on the local. The union won support only by increasing pay in tandem with shorter hours.

Until the number of baby-boomers in the workforce reached critical mass during the late-1960s, union efforts on health and safety were haphazard. Local meetings at Reimer's Co-op Refinery heard regular reports on fire hazards, first-aid measures, and precautions to be taken with chemicals, the local's minutebooks indicate. An industrial hygiene campaign was mounted on February 26, 1958: "lockers to be cleaned to get rid of mice and rats." Other zealous locals, like the ICW's Cyanamid plant in Welland, Ontario, assigned stewards to safety committees that inspected work practices and equipment as early as 1948, minutebooks show.

The lack of central direction from the labour movement shows through in the inconsistency of local agreements. Sometimes, workers' rights were threadbare. The 1956 contract with Fiberglass of Sarnia, for example, pledged the company to obey the law and the union to help enforce company safety rules. Workers at the atomic energy plant in Chalk River signed off their right to marry co-workers. That's how management handled the possibility of genetic damage, according to the September 19, 1955, *Union News*. A few locals won major advances. The 1955 agreement of Sask Power workers was 20 years ahead of its time. It spelled out that health and safety matters "shall be subject to negotiation between the Company and the Union."

In 1960 Reimer wanted to do something more systematic and sent out a health and safety questionnaire to OCAW members. The only worker's complaint that sticks in his mind came from a Brandon refinery. When snow from the roof melted on sunny days, the sidewalks got icy, a worker wrote back. Employers knew better. When Reimer asked to circulate the questionnaire on the shop-floor, some managers "got very uptight, felt I was going to sue them. They thought I must know something, when in fact I knew bugger all. Some called me in and told me there were toxic substances. That told me that we had to create a much greater awareness, that information on dangerous goods should not be kept away from the members, and that the union had not yet done its job."

That year Reimer directed all locals to set up union-management health and safety committees. If managers refused — as was their legal right until the 1970s — Reimer directed locals to set up their own committees. Then tragedy forced the pace of change.

In 1960, workers at the Phillips Petroleum gas plant in Fort St. John in northeast British Columbia told company engineers that a badly wired motor was throwing sparks. Engineers brushed the warning off. A short time later two men were killed in a kerosene explosion set off by the sparks. Workers wildcatted, and company president Kelly Gibson hit the panic button. He picked up Reimer in his corporate jet, flew him to Fort St. John, and the two negotiated a return to work. Management agreed to a joint health and safety committee. In the event of deadlock, Reimer and Gibson would arbitrate — something that no plant manager really wants, Reimer says. That clause in the agreement has never been used, and the plant has since won several safety awards. Reimer credits the awards to the union's first health and safety strike.

"We sort of grew into health and safety, not knowing a hell of a lot," Reimer says. He came across a sour-gas plant in Alberta where only one worker lived to retirement; the lone survivor died at 67. The union later found out that the hydrogen sulphide in sour gas was a nerve poison. It reminded Reimer of the "sudden death" that once haunted workers making explosives. Nitroglycerine caused a slowed-down heartbeat on the job, and workers suffered massive heart attacks as soon as they were away from the plant on holidays. Reimer decided it was time to tackle the poisons that get under workers' skin, no matter how carefully employees followed safety rules.

There was no one else to lead the way in Canada, least of all doctors. Medical associations "appear to be more concerned with fighting the introduction of socialized medicine than advocating and implementing" research into toxic chemicals, OCAW's Canadian Council charged in 1964. In 1968 Reimer asked Tony Mazzochi, legislative director for OCAW in the United States, for help.

"It has become clear that we have a more acute problem than many of us realize," Reimer wrote Mazzochi. He wanted his advice on workers' compensation claims and how to produce questionnaires to track down problems. "It will help formalize our approach and convince our local unions that serious action must be taken," he stated.

Mazzochi is the Benjamin Franklin of U.S. labour. The perennial odd man out in top U.S. union circles, he has Franklin's combination of wild hair, unkempt political thinking, and love of flying kites in stormy weather. A key figure in the merger of chemical and oil unions that formed OCAW south of the border, Mazzochi insisted on representing atomic workers. That way, he hoped, unions could gain hands-on leadership of efforts to tame the "peaceful atom."

Mazzochi made his first contacts with scientists in the 1950s, when he got workers in his home local in New Jersey to donate their children's baby teeth to scientists who used them to prove the bone damage created by nuclear weapons testing. Later he became the real-life mentor — terribly caricatured in the movie starring Meryl Streep — of Karen Silkwood, the union activist killed in 1974 before she could blow the whistle on unsafe practices in nuclear plants.

Mazzochi's modus operandi was to "expand the issue, expand the support base." He and the coalition OCAW built are credited with passage of the U.S. occupational safety and health act in 1970, the only major political victory of U.S. workers since the 1930s. "If you're worried about our getting our goddam foot in the door, we'll kick open the whole Pandora's Box," he threatened Shell executives during a bitter 1970s strike to win joint health and safety committees.

When he kicked open the box of industrial toxins, he showed the link between what went down in the workplace and what went up and out in the air, water, and food supply. OCAW had been making that connection since the 1960s, when it promoted Rachel Carson's *Silent Spring*, a landmark exposé of pesticide poisoning of water and birds. In 1970 Mazzochi was the principal labour organizer who worked with Hollywood celebrities to sponsor Earth Day demonstrations that put the environment movement at political centre stage. By 1971, as *Union News* reported in September of that year, consumer rights champion Ralph Nader was calling OCAW the foremost organization pressing for zero pollution.

Mazzochi's fast and loose style attracted the best and brightest of radical young doctors and scientists. Among these was Jeanne Stellman, who approached Mazzochi after hearing him at New York's Earth Day rally and agreed to write a manual for OCAW educationals.

The manual was turned into the book *Work Can Be dangerous To Your Health*, which became the standard text of the health and safety movement. OCAW challenged the very idea of "better living through chemistry." Stellman's index listed poisons common to oil and chemical industries -- 19 in fertilizer plants, 52 in paint factories, 34 in oil refineries, and 70 in plastics shops. The book pushed self-reliance and workplace struggle to open new frontiers of labour activism.

Workers needed the "right to know" and equal say in joint health and safety committees, Stellman said. Health and safety wasn't just another branch of union activity. It had to give direct power to the rank and file. Workers had to become experts, not rely on them. Stellman also lifted the curtain on workplace sources of mental ill health. Stress and monotony aren't the fault of individuals, she said. They're health and safety problems, caused by faulty work organization and bad management.

Mazzochi stressed the importance of unions linking up with scientists so workers could break the employers' stranglehold on information and take their own measure of chemical dangers. Thanks to the research and strategy pioneered by the U.S. section of the international union, Mazzochi says, "The Canadian section of OCAW's health and safety program was off the ground and running before there was any other group around in Canada." Reg Basken, who co-ordinated OCAW's health and safety program during the 1970s, says, "We were generous. We stole from everybody."

In April 1969 Reimer invited Mazzochi and a team of scientists to bring their road-show to Montreal for the union's annual Canadian Council meeting. Reimer scheduled the conference in Quebec to force attention on the stench of outdated plants concentrated there. But instead of focusing on unsightly pollution created by obsolete technology, delegates went after the unseen pollution created by state-of-the-art methods. I feel like the guy who says "for every silver lining, we'll provide a cloud," scientist Glenn Paulson apologized. He asked whether anyone at the meeting worked with PCBs and said they were trying to track its effects.

Tom Towler, Canadian representative on the union's international executive, told the meeting that he thought people were too easily led around by their nose. They equate pollution with stench, not with the quality of the world's oxygen supply, he said.

The problem, Paulson replied, is that people think oxygen extends to infinity. It doesn't, he said, and the oxygen supply is heating up. The air is fouled with carbon dioxide, the plant life that could absorb it is being cut down, and the trapped heat will upset the planet's climate, lead to meltdowns of polar icecaps and massive flooding, Paulson said, almost 20 years before "greenhouse effect" became a household term. The union's gloom and doom was two decades ahead of the time.

Delegates warmed to the sympathetic scientists and poured out their hearts in classic 1960s-style consciousness raising.

Of 140 workers in our box plant, we've had eight heart attacks and two cases of cancer in the past 16 months, George Walters said.

I've arranged to donate my body to science, said Ivan Hillier from the Dow plant in Sarnia, but "I don't want to give it too soon." Mercury is all over the factory, but workers can't even see the results from monthly urine tests to gauge its effects, Hillier said.

Managers called mercury a "respiratory irritant," not a deadly poison, but workers wildcatted two weeks ago to get the mess cleaned up, a Vancouver delegate said. "If they're unconcerned about pollution, you can bet your lives they're unconcerned about lives," he said.

In 1970 OCAW's public relations director, Ray Davidson, released *Peril on the Job*, a book of horrors based on testimony from chemical workers and scientists across the continent. Davidson prescribed strong union medicine to ward off the health crisis. Unions should become equals with managers on joint health and safety committees, he said. They should study and prevent hazards. Workers should refuse unsafe work. This was the new trinity of workers' rights in industry — the right to participate, to know, and to refuse.

While researching his book, Davidson was sent to meet two Uniroyal workers in Edmonton. A woman worker there got nervous when all her houseplants died, and she called Mazzochi. "She was like a Shell no-pest strip, killed every goddam plant in her house," Mazzochi said.

When Davidson was sitting in his Edmonton hotel lobby waiting to meet the Uniroyal workers he saw two men come in. They were immaculately scrubbed, wearing fresh clothes, and carrying a strong scent of aftershave. He recognized them instantly as the Uniroyal workers when the even stronger and "cloying smell of chlorophenols came into the room with them." They worked at manufacturing 2-4-5-T, the deadly Agent-Orange used by the U.S. air force to "defoliate" North Vietnam and poison the crops and soil that fed its peasant army.

The surge of energy from OCAW conferences and publications sparked action across the country. Cliff Hagen, chief steward of the Sask Power local, left the 1969 conference determined to find out more about the mercury used in Regina workshops to build meters. The union demanded an investigation by the health ministry, which found 20 workers contaminated, suffering from chronic headaches, memory losses, and violent mood changes. By 1971 the workers had won a cleanup, *Union News* reported. Exhaust fans were installed, safety equipment was provided, and hand-wash cleaning of mercury was discontinued.

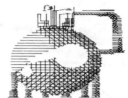

In Sarnia Ivan Hillier became one of Canada's first "whistleblowers."

An avid outdoorsman, Hillier had spent his childhood doing environmental rounds with his father, a district gamewarden for Cape Breton. They restocked lakes with fish from government hatcheries. They reintroduced beavers, previously driven to extinction on the island. As a teenager Hillier organized a petition campaign to turn a local marsh into a game sanctuary.

A typical damn-the-torpedoes personality from Glace Bay, Hillier was hired on at Dow's Sarnia plant in 1950 where he says he was "quickly branded as a combination of Jimmy Hoffa and Ivan the Terrible." CCF politics, fierce class consciousness, and hatred for the tyrannical style of the Miners' union were part of his folk heritage and led him to join OCAW in ousting the Mineworkers' District 50 at the plant in the mid-1950s. In the 1960s, when he was too old for Sarnia's senior hockey and baseball teams, he became president of the Dow local and the Sarnia labour council.

Hillier worked in the chlorine unit at Dow, where mercury was used to separate chlorine from salt. He saw "tons and tons" spill onto the plant floor and leak into the St. Clair River. He knew it was dangerous stuff. The company tested workers' urine every month to gauge the buildup of poisons but withheld the results from the workers. When a sympathetic office worker passed him the results Hillier formed a "vigilante group," located a contact in the University of British Columbia science department, and started researching the history and hazards of mercury. In 1962, he presented his findings to the executive council of the Ontario Federation of Labour.

In the 1960s Hillier worked with environmentalists, joined their study groups, and concluded, "Companies are in business to make money and they don't give a damn about research into environmental dangers." In 1970, according to Patricia Drummie's *History of Labour in Sarnia and Lambton County*, Hillier got the labour council to invite Pollution Probe to Sarnia. The two organizations prepared a scientific inventory of the damage caused by dumping mercury in the river.

The Ontario government, which admitted on April 2, 1970, that it had known about fish problems in the Sarnia area for a year, was on the spot. NDP leader Donald MacDonald, who'd been pressing the government to take action on mercury poisoning since 1964, finally got the public's attention. On April 2 he revealed a letter from Sarnia's Conservative MPP to Dow, asking company advice for a speech attacking MacDonald. The speech, the MPP said, "can go off on a tangent, just any tangent at all," and "without any great trouble take great pains to scorch Donald MacDonald."

On April 7 MacDonald asked the premier to investigate an anonymous brief detailing Dow's record of dumping 30,000 pounds of

This Award Goes

To Mr. I Hillier

Your Wonderful

I Thankyou for talking to us about mercury

from Tania Boyd

Apartment 17,
313 Indian Rd. S.,
Sarnia, Ontario.
N7T 3W6
1984 10 21

Dear Mr. Hillier,
You have alot of courage
to stand up to your Chemical
and say stop throwing all
that mercury into the river.
You deserved the Marguerite
Lind Vernon Weasley Award.
You're the kind of person
we need in Canada. If everybody
was like you and stood up to
such a big industry Canada
would be a more beautiful place
to live. You're a terrific guy.

Yours sincerly
Tina Cormier

891 Westbury Ct,
Sarnia, Ontario,
N7T 7R4
1991 10 21

Dear Mr. Hillier:
I think it is amazing
how you fought against that
big industry. I appreciate
what you did for our water.
I know how dangerous
mercury is to man. To me you
are the greatest.
Yours sincerly,
Chris Sauve

mercury a year in the St. Clair River. The premier refused to reply to an anonymous letter. On April 8 MacDonald announced that Ivan Hillier was prepared to go public and show company logbooks confirming that the company was tipped off whenever government inspectors came to investigate mercury spillage.

Rumours started to fly in the plant that Hillier was going to be fired. He got late-night phone calls denouncing him for risking jobs in the area. Threats were made against his children. The union was accused of opposing the smell of money. "I didn't want the union to get screwed up, but I couldn't live with myself if I didn't follow my principles," Hillier says. On one trip to Toronto, his flight back to Sarnia was held up by bad weather. When he hadn't returned on time, a police manhunt began. While plans were being laid to dredge the St. Clair River to look for Hillier in a pair of cement shoes, he was sipping his beer at the Royal York Hotel, waiting for the train back to Sarnia.

On April 30 Ontario banned fishing in areas contaminated by mercury. On the next day commercial fishing was shut down. A year later, March 15, 1971, Ontario sued Dow for $25 million for wiping out commercial fishing and threatened to sue for another $10 million if Dow didn't clean up the mercury sealed in river beds. Hillier called it a show trial. The government was so involved in company wrongdoing that it won't be able to establish company guilt, he said. The company based its defence on its record of co-operation with government. On June 9, 1978, the government settled out of court for $400,000, to be split between the province and fishers. Over the next decade Dow was forced to spend $40 million to switch to new processes that eliminated mercury pollution.

In Alberta Neil Reimer gained a reputation as an environmental crusader during his years as provincial NDP leader in the mid-1960s. He knew from chemical workers at Two Hills that sulphuric acid was being dumped into the headwaters of the Saskatchewan River, and he knew from NDP contacts downstream in Saskatchewan that the water supply was endangered. He made a splash with the issue by putting table water from North Battleford on legislative desks and inviting Socred politicians to take a sip.

When he resigned as NDP leader and returned to full-time union work, Reimer tried to overcome public apathy about workplace pollution by stressing the link with impure food, water, and air. In the midst of a public scare over food additives in 1970 he told the *Edmonton Journal* that "If some of the new products on the market today are harming the consumers, just think about the harm they are doing to the workers who are producing them day in and day out."

Two weeks later, on May 25, 1970, the OCAW convention in

Edmonton directed bargaining teams to table union demands for pollution control. It also called on politicians to withold factory building permits until union-government inspectors were satisfied with health, safety, and environment standards.

From 1971 to 1977 Reimer chaired the Alberta Federation of Labour's environment committee, which handled the training of workplace health and safety committees. The Alberta Federation declared the environment its major theme of 1971, and the committee produced a slick booklet, *What is Labor's Stake in Environmental Pollution?* It called the workplace the "prime source of pollution."

For Reimer this approach was common sense, simply a matter of putting two and two together. The idea of splitting up the issues, he says, "came from conservative governments that wanted to treat the workplace differently from the environment. The labour movement later went along with that split, and allowed the opposition to government policies to be divided. I resisted that separation. It's always been the hallmark of our union that we kept them together."

In 1973, as Reimer boarded a plane from Vancouver to Tokyo, he was called into the pilots' cockpit to take an emergency call from the Alberta premier, Peter Lougheed. The plane was held hostage until Reimer agreed to join a provincial commission on health and safety.

Chaired by Fred Gale, an engineer and general manager of a public utility, the commission included three corporate safety managers, two unionists, and Dr. Charles Varvis, a specialist in internal medicine. They got down to work in November 1973, toured the province to hear 76 briefs, and presented their report early in 1975.

Though outnumbered by managers, Reimer had an agenda. He burnt the midnight oil, wrote rough drafts that commission members criticized, then worked with Varvis on rewrites. As in cards so it is in rough drafts— the advantage is always with the house. The writer defines the approach and gets to keep everything that isn't challenged. The final report called for a total overhaul of health and safety administration.

Health and safety needed a home of its own, the commission said. Until then, it was orphaned among six provincial ministries and departments, on the first-serve list of none. The commission's proposed "one window" approach, now accepted wisdom, forced responsibility on one minister, made it harder to pass the buck, provided a multidisciplinary setting for staff experts, and highlighted a clear identity and high profile that would make health and safety a new provincial priority. Likewise, the commission wanted industrial medicine, one of the most neglected areas of health care, treated as a specialty in a medical school and hospital.

"There must be an entirely new approach" to workplace regulation, the commission said. "Emphasis must be placed on the involvement of both the worker and the employer in the maintenance of a safe and healthy environment at work." The commission favoured joint labour-management health and safety committees with the right to know about workplace chemical hazards and the mandate to develop educational and preventive programs.

Later in the 1970s, the principles of one-window administration and the legislated right to know and participate became standard fare in government task forces and regulations dealing with health and safety. One Gale commission recommendation never did get picked up. It's not enough to have joint committees on health and safety, the Alberta report said. Even more critical were the involvement of workers, their ability to make decisions on the job, and their feeling of acceptance and responsibility.

In particular, the commission said, "The monotony of many jobs may produce apathy and boredom which could result in accidents." This was the old myth of the careless worker turned on its head. Careless workers were really dying of boredom. To correct that, something had to be done about job satisfaction. "The Department of Occupational Safety and Health, employers and the workers' organizations should be expected to actively pursue improvement in the quality of working life," the commission said.

Quality of Working Life, a North American translation of Norwegian workplace democracy experiments, became one of the buzzwords of labour relations experts in the mid-1970s. But health and safety activists rarely linked their concerns to questions about job satisfaction and workplace democracy. The Gale Commission remains the exception in its focus on the need to empower workers as individuals, in its linkage of mental and physical health at work.

"The nucleus of the thought," Reimer says, came from the commission's tour through Europe. They met "Chip" Levinson, the Canadian-born leader of the world-wide oil and chemical union federation (the definitively non-CIA federation), who introduced them to World Health Organization officials then flirting with advanced ideas in industrial democracy.

In response, Premier Lougheed unified all workplace health and safety agencies under one ministry. In 1976 he gave workers the right to refuse unsafe work. He also matched the corporate donations Reimer raised to launch a chair of occupational medicine at the University of Alberta. The school publishes the *Canadian Journal of Occupational Medicine*, which promotes research on Canadian subjects neglected by international journals.

Otherwise, Gil Reschenthaler writes in *Occupational Health and Safety in Canada: The Economics and Three Case Studies*, "The expected reform has not materialized." The regulation of employers petered out.

217

In 1977 there were only 45 inspectors, who reported 50 deaths and 9,888 violations of legal standards. Though repeat violations were classified as "endemic," only two companies were prosecuted. The average fine was $200. That year, only 47 workplaces — none in construction — had health and safety committees set up under the act. Workers' input was "consultative." In 1979 the government gave up the pretence of enforcement and declared the committees voluntary. The government gave in to employers who feared any workplace organization that might lead to unionization, Reimer says.

But the commission's vision of health and safety reform is second in importance only to the work pioneered by the Saskatchewan NDP in 1972. Reimer, his work honoured more in the breach than in the observance, received the Alberta Award for Excellence for his contribution to the findings.

Government inaction forced unions to bargain for health and safety rights. National bargaining allowed the union to place health and safety high on its agenda. By contrast, unions that scramble to settle deals on a strictly local basis rarely have the clout, energy, focus, or vision to deal with such thorny problems. There's too much pressure to get to the hardpan, to drop a troublesome issue that stands in the way of agreement, to deal with it as a demand to be met by government in the fullness of time.

OCAW got the attention of oil companies in September 1969, when it led off the conciliation hearings in the 1969 B.C.-based strike with health and safety demands. Public hearings were an ideal environment to raise such issues but the companies cried foul. They complained that these new issues were being raised only for their embarrassment value. The union had pulled a similar stunt, the companies believed, when the strike had begun that spring. Chief negotiator Buck Philp denounced strikebreaking supervisors for sloppy workmanship that allowed tons of aluminum dust to escape into the atmosphere. Philp called on the city of Vancouver to shut the polluters down.

Union demands in 1969 were a lucky shot in the dark. In 1971 the ideas sunk in when Reg Basken, Reimer's assistant, was appointed the union's health and safety co-ordinator, a signal that the issue was of prime importance and that resources would be devoted to it. One of Basken's jobs was to keep local health and safety committees up to date with the latest scientific information on chemical hazards. The new emphasis led to an intellectual revolution, he says. "Before that, we spent almost all our time on safety issues."

In 1977 a national bargaining conference for oil workers outlined a three-part health and safety program. Workers demanded the names of all substances used in production, and they asked for management co-operation to gather research on these substances. They also wanted equal union input into precautions for handling the substances. The demands were accepted in 1978, and the victory was hailed by the union as "the single most significant improvement in bargained health and safety language in Canada."

Nine years later, in 1986, federal and provincial governments co-operated to establish WHMIS (Workplace Hazardous Materials Information System), and called it a pathbreaking initiative. In fact, WHMIS only requires suppliers to provide names of industrial substances and any information they have gathered. There's no requirement for further research, or for union input into research and proper precautions. In short, there's no shift in power from managers to workers.

In 1988 oil workers took their decade-old victory one step further. Petro-Canada agreed to check off three cents an hour for every worker to finance union-sponsored education and training in the health and safety field. Since then, all national bargaining programs of the Energy and Chemical Workers have pushed contributions to the union centre, which Russ Pratt, co-ordinator of the training program, says is "a baby step to equalize resources."

According to Pratt, "The biggest problem with union leaders is we don't know how to plan. In the sixties and seventies we drove change — pensions, medicare, inflation protection — and the companies reacted to us. In the eighties management drove and we reacted. There's two reasons for that. The labour movement is complacent, trains its stewards in the same old way. And before, we had all the information. We did grievances and negotiations all day, every day. They couldn't keep up with us. Today, they have computers, they're hooked up to instant access to all the information. We don't."

"The only way to make change," Pratt says, "is to get people to come to grips with why they think and how they think, why people stop being careful, why most workers prefer not to know about the chemical hazards they work with."

That strategy, sometimes called "empowerment," is at odds with the technocratic approach of government and industry. That approach, exemplified by WHMIS, focuses on training, not education, and on minimal rules, not optimal ways of setting rules. It assumes that it's the information gap, not the power gap, that has to be closed.

The technocratic approach represents a new paradigm for health and safety, at odds with the model developed by energy and chemical workers. The differences between these strategies will define the health and safety debate in Canada throughout the 1990s.

KILLING TIME

The ancient Greeks called it asbestos, "inextinguishable." Its fibres couldn't be beat or burnt, and industrial society's demand for the all-purpose mineral was unquenchable. World sales doubled from 1960 to 1973, and Canada was a leading producer. Asbestos was put to over three thousand uses: to insulate walls and clothing, to line friction points of machines, to reinforce cement and tiles.

Bad news didn't travel as fast. The long and curly fibres the lungs can't clear are five-millionths of a metre. No one drops dead after breathing them. They are "timebombs" that can take 20 years to click off.

Still, Pliny the Younger observed the trail of death among asbestos weavers in 100 A.D. The folk wisdom in Quebec factory laws led to controls on asbestos in the 1890s. A medical researcher amassed statistics on French textile workers in 1906. Canada's *Labour Gazette* reported on dangers in 1912. Life insurance companies were warned about asbestos workers in 1918. A 1930 British royal commission linked asbestos to lung diseases. Ontario's Workmen's Compensation Board allowed asbestos claims after 1942. Five years later, the labour ministry had strict guidelines against dust.

As the dust cleared and workers lived longer, asbestos was linked to other diseases. In the 1950s and 1960s, British and U.S. researchers found that asbestos workers had six and ten times the average lung-cancer rates. Mesotheliomia, cancer of the lung linings, was also linked to asbestos. So were arthritis, right-sided heart attacks, and pneumonia.

Yet Ontario's 1984 royal commission on asbestos blamed the slow workings of science for delayed government reaction. "As with asbestosis, so with lung cancer, the path from initial identification to full recognition of the association has taken decades," the commission stated. On release of its report, the Construction Safety Association, which is financed by the Workers' Compensation Board, ran TV ads repeating "If only he had known" alongside pictures of an asbestos victim.

If only industry and government had told. Johns-Manville knew when it built its Scarborough plant in 1948. In 1949 the company's Canadian medical director found X-ray evidence

220

DUST TO DUST

of asbestosis. "But as long as the man is not disabled," the doctor wrote head office, "he should not be told of his condition so that he can live and work in peace and the Company can benefit by his many years of experience."

It's the connection of government, corporate, and medical neglect that was slow in coming. Once asbestos had burned that connection into the minds of unionists, ideas that had been the preserve of a small minority became commonplace. Health and safety went together like Johns-Manville and asbestos. So did the right to know about workplace dangers and the right to refuse unsafe work. So did the responsibility of unions to take up a new issue to protect their members.

Alf Glaser just wanted to finish his 30 years at Johns-Manville and get out on full pension. He'd been active all his life, kept up a brisk pace at work, fished and hunted, pitched in to help neighbours build their own houses in the eastern suburbs of Toronto. He didn't smoke and had stayed firmly on the wagon after a brief problem with booze.

He almost made it. In 1979, a year from retirement, the company doctor told him he was in A-1 health, "nothing to worry about." A few months later he came down with terrible stomach pains. He had ulcers, he should diet, the doctor said. For three months he lived on milk and porridge. His stomach stayed bloated, and he caught pneumonia.

The doctor at the Ajax hospital broke the news. He had cancer, it was too late to operate, and he had only months to live. In September 1980, under heavy doses of morphine to suppress his pain — "It's just like a knife cutting into you, and then I start vomiting" — he gave his life history to researchers with the Canadian Centre for Occupational Health and Safety. They wanted to present it to the Royal Commission on Asbestos.

Alf Glaser was born in Russia in 1924, came to Canada in 1950, and started at Johns-Manville on June 14 that year. He made moulds for asbestos-cement pipe. "Inside the dust was flying all over the place. You had no mask on or anything," he said. "There was nothing to enjoy. You had to work to make money, never mind enjoy it. We didn't have anything. I didn't have a place to stay, nothing," he said. He was never warned about the dangers of asbestos.

In 1959, his hearing gone from the clatter of the moulding machine, Glaser was transferred to the unloading department, where he and three other men hauled hundred-pound bags of asbestos from boxcars. They all worked "on bonus."

"All we were trying was to unload as much as we can, to make bonus." Glaser said. "The faster you worked, the faster you walk and everything. It's just blowing around us. You couldn't see the two other guys on the other side."

He asked for another transfer. "I couldn't take it any more in there. I was bringing up always. I didn't think there was anything wrong with me, so I just went to the manager and I asked him if I could change jobs. I said I can't take the dust, something's wrong." Glaser never thought of quitting, feeling that he had no choice in the matter because he spoke only one language. "I couldn't even speak English at that time. When I found a job, I had to hang on."

In the mid-1970s, Glaser heard Ontario NDP leader Stephen Lewis speak about the emerging health disaster at Johns-Manville. The news scared him but by then he had put in so many years he had seniority. "Your good years are past, you can't jump from a place, one to another," he said. "I thought I was gonna pull through. I would take the retirement and get out, just as long as I'm still well enough. Well, it caught up before I got out of there."

Glaser's wife quit her job to stand by him in the hospital. His family came every day. Fellow workers dropped in, and the union sent him a little cash. But, his wife said, "Not one representative from the company went down to see him or sent a card. You work for somebody for thirty years, you put your whole effort and your whole life into a company like that. So I figure a worker is at least worth a card."

Glaser's brother, Gus, and their uncle also worked at Manville. They too got sick from asbestos. "My brother-in-law died like an animal, just like an animal," Gus's wife Betty Glaser told the royal commission in 1981. The same thing happened to her uncle, she said. At the time Gus was still alive, hospitalized since 1971, living off four tanks of oxygen a week, unable to prove to the Workers' Compensation Board that he had a respiratory disease. The WCB said he had arthritis and pensioned him off at $135 a month.

From the mid-1970s on, a widow's watch opened the eyes of the country to a new breed of workplace accident. Johns-Manville put Scarborough on the map of world-class health disasters, the Royal Commission on Asbestos said. In a workforce of 714, the WCB compensated 68 deaths and 113 disabilities. It did not compensate 71 workers who died of asbestos-related diseases. A follow-up study by the Ontario ministry of labour found that plant workers, including those who stayed at Johns-Manville only a short time, had five times the normal lung-cancer rate.

The era of catastrophe, of Three Mile Island, of Chernobyl, of Bhopol, of AIDS, was about to begin. Tiny specs of dust changed the way people saw governments, scientists, the environment, and workplace health and safety. Old sayings about "out of sight, out of mind" or "you eat a peck of dirt before you die" lost currency. Government regulators and scientists lost the public's trust. Better watchdogs were needed for invisible poisons. These lessons were learned the hard way. The Energy and Chemical Workers learned the hardest. The International Chemical Workers, which merged with ECW in 1980, represented workers at Johns-Manville.

The remains of loggers, miners, deep-ocean oil riggers, and fishers rest in ghost towns, cemeteries of the boom and bust resource frontier. Crashes, explosions, and storms have always been a way of death in these communities. The remains of steel and textile workers, racked by cancer, silicosis, and brown lung, rest in single-industry towns. The remains of construction workers lie crushed in the inner-city core, beneath highrises and subways that met their deadlines.

The remains of Manville workers rest in Scarborough and Ajax, in cemeteries of the suburban dream. In the flush of post-war prosperity, workers fled downtown pollution and overcrowding for suburbia. The Johns-Manville disaster is a study of the suburban mind, of the great escape from Depression memories, of the blind faith in how personal security and family happiness could be found outside sound social policy. Manville was the mind-gulag of "Scarberia," where see no evil, hear no evil, say no evil, leave-things-well-enough-alone, became a way of life and death.

Like most suburbs, Scarborough rolled out the red carpet for industries that could rescue it from near-bankruptcy. Homeowners alone couldn't carry the tax burden. After the war the township set aside a "Golden Mile" to lure companies with cheap land and tax breaks, no questions asked. Johns-Manville set up shop in 1948 in the farmlands and wilds of the Rouge Valley.

Although neighbours were upset by the noise and by what smelled like linseed oil, writes John Spillsbury in his *Facts and Folklore on Scarborough History*, they thought the problem would blow over when the company built a 200-foot-high stack. Local kids played hide and seek in the asbestos heaps and used asbestos to line baseball fields in the schoolyard just down the way.

Manville was an ideal employer for new immigrants. The work was steady, the pay was decent, and the overtime — often 20 or 30 extra hours a week — paid the mortgage. Some workers who filed compensation claims after ten years at the plant had 16 years' worth of time credits. They had but one life to give, and they gave it to the company and their families.

You have shown "patience, understanding, co-operation and courage," the company told men who made it to the 25-year club, the model employees and citizens in the era of suburbia. "You have not insisted that the world owes you a living. You simply have wanted the world to give you an opportunity to live and make your own careers. Unlike some, you have not attempted to destroy what others have built so that they might profit thereby. Rather you have built for yourself." Shoulder to the wheel, play it safe and steady, look after yourself, don't ask where the profits go. Manville workers were the success stories of this way of life. The company lowered its flag for a day when each one died.

"I don't need the very best, because I have the very best," Manville worker John Dodds used to tell his wife, Odette. "A wife, two good kids I love very much, the wish of a king, a home, our home, a little icing on our cake, a few pennies for the rainy days and early retirement, a brighter promising future."

In 1978 John Dodds died in agony and his family was left without a penny in compensation. "As you grow older, you grow wiser. The worst part of it is, you grow old too fast and wise too late," Odette Dodds told the royal commission.

The dreams of the Dodds were simple and sweet, but like all dreams they had their price. "Greed becomes involved, you know," long-time local union president Charlie Nielson told investigators from the Canadian Occupational Health and Safety Centre. "That's a bigger disease than anything, I find. These guys get this bloody greed in them and, you know, they can't see past their bloody nose. They're quite willing to give up their life for that."

Suburbia was supposed to be about leaving the cares of work behind, finding joy in family and home. Johns-Manville taught a generation that's not what work's all about.

A good dunking in asbestos was a Manville ritual, a rite of passage. Bruce Machin, company vice-president, told the commission about initiation procedures in place when he started on the floor in 1963. He said that when he was taken on the rounds by the union president, "I found out how to work a dust collector by putting my head inside the hopper bin and having somebody turn on the shaker and bury me in the hopper with asbestos." At the time this kind of foolplay was "thought to be quite humorous."

Workers couldn't see the asbestos fibres, but they saw the dust. Throughout the 1950s and 1960s, Nick Carrigan worked on a crew that shook up bags of blue asbestos for machine operators. The dust was so thick, he told the commission, "You couldn't see the guy sitting beside you. And while this was going on, we sat down and we ate our lunch in it." They wore no masks or protective equipment. Of seven men on his crew, he was the only one alive to testify at the commission.

The company's first ventilation system made matters worse. Dwight Oland started at Manville in 1950. Many machines weren't hooked up to vents, he told the commission. When the main vent got clogged, he said, "Men were forced to crawl in through it on their bellies with scoops and buckets, dropping the dust contents onto the floor." To save on heating bills the company adjusted its ventilation system so, as worker Ed Cauchi told the commission, "the dust control system was shooting the recirculating air inside the building." This multiplied the chances of breathing in fibres.

Dwight Oland told the commission, "Employees continually complained about the dust, even though many of them were unaware of the danger of inhaling the dust. It was a nuisance at least. It was on their clothes, it was in their lunches. It was in their coffee cups. It was everywhere. It collected in our autos and it was taken home on their clothes." Foremen shrugged off complaints, said a little dirt never hurt anyone, and if they didn't like it they could always work elsewhere.

Workers were left to believe what they wanted. In 1979 a government doctor told Oland he had nothing to worry about.

"I thanked him for putting my mind at ease, and said it was the best news I had heard in years," Oland said. By the time he testified to the commission, he had suffered a heart attack and arthritis, both common to long-time Manville workers. He also found that X-rays documented problems as early as 1968, though the results had been withheld from him.

A maintenance welder, Joe Pagnello, started at Manville in 1961, cleaning out the ducts loaded with asbestos. Pagnello said, "We knew there were people got sick from asbestos" but "we didn't take it that serious, unfortunately." Around 1970 he had trouble breathing and found he couldn't keep up in square-dancing. Like most people in the pre-medicare generation, he hoped the problem would go away and didn't consult a doctor. Doctors were for really serious illnesses that wouldn't go away. In 1977 the company nurse knocked at the cafeteria window, waved Pagnello over, and introduced him to a man from the Workmen's Compensation Board who told Pagnello he was eligible for retraining. That was how he first learned something was wrong.

"It just hit me like a ton of bricks. Of course, by this time we know how serious these things are, and everybody is just hoping that they aren't one of the unlucky ones," he told the commission. He couldn't find work after the plant shut down in 1980. No one will hire an ex-Manville worker, he said.

Despite parched throats, short breath, and a fifth sense that so much dust had to be dangerous, the truth about asbestos came as a shock to the Manville workers. They were like lambs led to the slaughter. Although the company knew about the dangers — shown beyond doubt in Paul Brodeur's books *Expendable Americans* and *Outrageous Misconduct* — it issued no warnings.

Ed Cauchi, later the leader of Asbestos Victims of Ontario, started at the plant in 1953. Workers "were never, ever told the dangers of working with asbestos, until 1974-75," he told the commission. "Until that time, the only thing discussed about safety with the worker, was make sure this is the John-Manville Safe Work manual," he said. "They talk about safety shoes, gloves, safety glasses, but don't talk about respirators whatsoever."

Long-time local union president Charlie Nielson said that for years the company doctor, who doubled as family physician for many workers, had "told the guys there's nothing wrong with you, it won't do you any harm, go back to your job, you've nothing to worry about." The doctor would tell them, " 'You'll live longer than I will....' And these people are all dead and he's living, and that's a shame," Nielson said.

According to Nielson, when a small band of unionists raised an alarm, the company pooh-poohed their fears, said asbestos was just "nuisance dust." Nielson said, "The guy on the floor, he'll believe the company before he believes the union. That's the way it is. If the company says that won't do you any harm and I come out with a piece of paper and say that's going to kill you, that's bullshit. Just as plain as that."

Dwight Oland, a shop steward and member of the local's compensation and environment committees, told the commission he has been trying to "get the company to lessen the amount of dust, to teach and maintain good housekeeping" since 1965. It "had about the same effect as trying to bore a hole through a brick wall with a pencil eraser," he said.

The company ignored its own research. In September 1970 *Environmental Science and Technology* magazine reported on Manville research showing the need to wet down asbestos to prevent airborne dust. Such simple precautions weren't followed at the Scarborough operation until 1975. That's also when masks were first made available. They weren't made compulsory until 1977, the same year separate lockers for street and work clothes were installed. Until then workers carried asbestos dust on their workclothes back home to be washed. That year the company sent a doctor and senior executive to warn workers against smoking and sponsored a smoke-enders course.

Three government ministries — labour, health, and environment — were responsible for asbestos regulation. None took initiatives on their own. They all refused to back up unionists who called for changes. Ontario had guidelines against excessive asbestos dust as of 1948. Yet a ministry of labour study by Dr. Murray Finklestein admits no fibre counts were taken until 1969, and no reliable ones until 1973. The province's occupational health branch didn't test workers until the 1960s. Results were withheld from workers, and no educational campaign was conducted for personal physicians. When Manville workers got sick, they were sent on wild-goose chases by family doctors who suspected ulcers and arthritis. Few of these workers had more than three months to live when they discovered the truth.

"This is why I said our biggest problem always was, and still is" the health branch's chest clinic, Ed Cauchi told the commission. In 1975 Cauchi tried to involve the Scarborough city council, which was responsible for death certificates, in a clean-up campaign. The council's deputy medical officer accused Cauchi of creating a false panic.

Although the company was allowed to flaunt the law, workers were held to strict account. If they couldn't prove their illness was solely due to the workplace, they got no WCB pension.

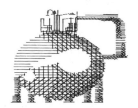

If Johns-Manville is a study of government and company fail-safes that failed, it's also a study of human and union fail-safes that failed.

Neither workers nor the union had the know-how, power, and resources to limit asbestos damage. But their slow reaction time remains a mystery. The ICW didn't have blind faith in company goodwill. It had a medical doctor on staff and used him as a drawing card in organizing campaigns, as proof of a modern union that cared. Manville workers all felt sickened by asbestos dust long before they got sick to death. The dreaded news came as a shock only because they ignored their own gut instincts and early warning signs. Something besides ignorance kept them from getting the problem off their chests.

To this day, workplace health and safety efforts centre on external agencies that regulate and train. Little attention is given to the deeper problems of the human mind or the negative psychological conditioning of the workplace.

In his 1968 study *Collective Bargaining in the Canadian Chemical Industry*, George Eaton found that health and safety issues were "not really matters of grave concern to the unions, but rather items to be used in bargaining for wages." That is a view shared by the union staffers who serviced the Manville local.

Ken Rogers, who later joined the Ontario Workers Health and Safety Centre, told the local in the late-1960s about Dr. Irving Selikoff's 1964 research on asbestos and cancer. He wanted to bargain for the right to see company health records. The local wasn't keen. "We were all like teenagers with health and safety. We just believed in the invulnerability of the human being, the `it-can't-happen-to-me' syndrome," Rogers says. "They were not really interested in getting information that showed they were getting killed. Mentally, all the shutters slammed down."

When Rogers returned to the local in the late 1970s to urge a breakaway to the Canadian Chemical Workers, "The son of one of the guys who'd died attacked me for the way I badmouthed the company, said they'd treated his dad right 'till he died. `They killed your dad,' I said."

Workers wanted the union to mind its own business, Rogers says. They were "more inclined to negotiate better sick pay than resolve the problem. It wasn't our role to negotiate the hazards. Our role was to negotiate compensation for the hazards."

In the jargon of Canada's industrial relations system, unions and managers are called "the parties." But management is the permanent government and unions are the perpetual opposition. Unions are suposed to be a loyal opposition, not one that bites the hand that feeds. Their job is to ask "what's the damage?" and then put some order to it, set a price on it. Unions complain about approaches to health and safety that focus on protective equipment, such as masks or earplugs. They talk about "engineering the problem out at source," getting rid of dust and noise. But as long as they accept their stilted role, they too are reduced to a piece of protective equipment.

Russ Pratt, a former ICW rep who co-ordinates health and safety programs for ECW, says that workers practise avoidance by not taking charge. "It's like people who won't go to the doctor when they have a sore. They don't want to find out, and they don't want to take corrective action," he says. "Do you want to live if you lose your legs? Do you want to die slow and have a job, or live long and not work?"

Why didn't the ICW doctor come to the plant and confront the problem? "I asked myself that a hundred times," Pratt says. "The workers wanted to avoid the truth, and we helped them avoid it."

"If the guys admitted the job was killing them, they'd have to quit," says Bill Adams, an ICW rep in the early 1970s. "So all we got was grievances on overtime." Manville was a U.S. company with a job classification plan that kept workers fighting among themselves. Its skilled tradesmen were usually from Britain, while labourers came from Eastern and Southern Europe. Many workers joined the bargaining team just to get a better rate for their own group, Adams says.

Workers got back at the company by "fucking the dog so they could get more overtime," Adams says. "In the back of their minds, they knew they had to store up money for their families. And the company was always prepared to feed the men more money rather than clean up the plant." Adams says when he asked for support from Canadian union director Tom Sloan, a former Manville employee, Sloan snorted, "That dust is nothing. You should have seen it when I was there."

In July 1971, a CBC television program blamed asbestos for four deaths and twelve disabilities at Manville. The program, together with the legislative debate it sparked, drew public attention to a bitter strike at Manville, which featured company use of professional strikebreakers. Strikers raised the problem of dust levels, which the health ministry claimed were acceptable. The strike was settled without resolution of health issues.

Adams credits his successor, Bob Stewart, for "spearheading the issue singlehandedly. He took it on as a personal cause." By the time Stewart arrived the death toll was mounting exponentially. "There wasn't hardly a union meeting that went by where there wasn't a minute's silence for one of the members who had died prematurely," Stewart says. "And that's really pathetic — to go visit a guy who's a robust healthy man at one time, and then six months later he's just a bag of bones lying on his deathbed."

In 1972 Stewart tried to co-ordinate all unions bargaining with the Manville chain. He organized a conference in Illinois to hear Dr. Irving Selikoff and played the doctor's speech describing cancer as a political disease caused by governments and corporations to local meetings. The union focused its attention on the sick, most of whom were in desperate straits because of the Workmen's Compensation Board. The Board worked from a "meat-chart" that gave awards based on the part of the body that was hurt. A hand was worth so much, an eye worth so much. Lungs were worth anywhere from 15 to 40 per cent of an injured worker's former salary, little solace for labourers who couldn't work at all without a strong pair of lungs.

Stewart got the CBC to televise a group of Manville workers who couldn't walk across a backyard. The WCB gave them a 15 per cent disability pension. Manville workers with cancer weren't so lucky. Unless they had worked at the plant for 20 years, the WCB awarded them nothing. The royal commission later described such guidelines as "not commensurate with the importance of the equity goal they are meant to serve."

Dan Ublansky, the ICW's staff lawyer, represented a dozen asbestos victims who appealed Board decisions. "It was tough. What we were confronted with was desperate people in their fifties and early sixties who knew their lives were going to be short and who just wanted to get things in order for their families," he says. "They were bitter about a system that made them fight for every last penny after they got injured at work. It was like talking to a wall. The Board had its way of assessing a disability and the workers didn't meet their standards." Ublansky didn't have much chance of winning. At that time, appeals against Board decisions were heard by another level of officials at the Board. The royal commission later described that system as "arbitrary and capricious, lacking in top-down direction and procedural fairness."

Ed Cauchi, chair of the local's health and safety committee after 1973, pushed the government to release chest X-rays to workers and pushed the company to vacuum out dust and provide separate lockers for work and street clothes. In 1976 he got the WCB to provide retraining programs so workers could be placed in less strenuous jobs. Charlie Nielson, the local union president, faced intolerable frustration. He had the unenviable job of harassing fellow workers into following precautions. He begged the company to punish workers who refused to wear masks. "We were caught in the middle between the company and the guy. We were the rotten so and so's. We weren't getting any encouragement from the company, they were wanting to keep us way out of there. And this is where I fault the company."

To force the government's hand on testing, Nielson became a lay doctor and got all workers to bring him a bottle of coughed-up spit. As for the WCB, the only answer, he told researchers from the Centre for Occupational Health and Safety, was "to go down there with a gun and shoot all of them."

Nielson also had to take on his own union. The International paid only lip service to issues of health and safety and the Canadian director of ICW accused him of making too much noise on the subject, he said. Nielson fought for an alliance of all unions that faced employers using asbestos but couldn't find any takers. "That's a bloody shame," Nielson said, "because the only way you're going to compete with the multinationals is to get together. We don't stand a chance in Canada the way we are."

In 1977, four hundred Manville workers petitioned the ministry of labour to investigate company negligence in not telling workers about asbestos dangers. When the government wouldn't sue, Neilson went to a lawyer. The International cut off his funds. That was the final straw. Led by Neilson, the Manville local voted 358 to 73 to dump the ICW in favour of an all-Canadian union that would seek to unify all chemical workers into one union. Bob Stewart and the former ICW staffers favoured joining the new union, the Canadian Chemical Workers.

At first Stewart wasn't sure how to solve the asbestos problem. He was open minded, he told *The Globe and Mail* on March 5, 1977. Maybe the plant could be made safe, maybe it had to be shut down, though Stewart tended to support British trade unionists who felt there was no safe level. It's a tough decision, he said, to call for a ban that throws union members out of a job. The dilemma was that the union membership didn't necessarily share the concerns that placed health over jobs. But Stewart stuck by his principles and won the members over. In 1980 a strike put Johns-Manville out of business.

The asbestos health crisis was heading out of control. "There is a total war on between the workers, the Compensation Board and the government," Charlie Nielson told the founding meeting of the Asbestos Victims of Ontario in 1980. Other unions, especially the

Autoworkers representing Bendix brake workers in Windsor, were moving on the issue. The public was feeling anxious too, worried about asbestos fibers floating about in buildings and in the air and water supply.

Ontario's Conservative government moved to shake the dust loose from its own record. In spring 1980 it set up a royal commission to investigate all aspects of the asbestos controversy. The commissioners appeared to be independent, fair-minded experts — safe people who could be trusted to take the problem in hand and hold it there until the public turned to other matters. Chief commissioner Stefan Dupre was a political scientist with several publications on science and fiscal policy to his credit, some experience in labour relations, and a solid record of conservative thought. The commission lawyer, John Laskin, was sired by Canada's most progressive judge, but built his own career on different principles. Later he represented the far-right National Citizens' Coalition in a case against union shops and co-authored a 1987 report that accused unions of using health and safety to promote radicalism.

The commission's hearings became "a lightning rod for the anger and the pain that had befallen the sick and the survivors," says Stefan Dupre. Televised hearings helped Manville widows touch the hearts of the country. Some of these women came at the behest of the Asbestos Victims of Ontario. Some came at the request of their husbands. Some came on the spur of the moment, inspired by the courage of others who broke the vow of quiet desperation. Almost all identified themselves by their husbands' first and last names. They blasted the company, the government, and the medical profession.

Betty Glaser came at the request of her husband Gus. The WCB had denied his request for asbestosis compensation, claiming he had arthritis. Betty Glaser told the commission: "My husband said to me, `look, I know I'm dying. I don't want you to have to go through what Odette [Dodds] or many other women are, not receiving a penny.' So now I have to fight.... A criminal kills a person, goes to court, he gets somebody — a defender — paid by the country. But we don't. Not one of us get any help, except that we help each other."

Jean Roseman was left with six children when her husband died at age 53. He had worked at Manville for 12 years. Taking overtime into account, the WCB granted him 17 years of exposure, three years below the cutoff to qualify for survivor benefits. "I did not only lose my husband," Roseman said. "I lost all my hope and faith and trust in people, in doctors and the government." She came "to shout it from the mountaintop: have mercy ... please have mercy on us."

Frances Day's husband Gord died after working 26 years at Manville. Whenever co-workers were diagnosed for asbestosis, he would say it was too late to quit, he'd lose his chance for compensation. He died in 1974. Frances Day got full compensation.

Mrs. William Barton watched her husband die in 1978, after 28 years with the company. "Three weeks before his death he was in such pain that he asked that I bring a knife into the hospital so that he could

231

end his life," she said. In 1967 his weight had dropped to 89 pounds and his doctor told him his lungs were full of garbage. He booked off sick for 26 months. In 1974 the WCB gave him a 10 per cent disability pension, later increased to 20 and then 50 per cent. Because he died of a complication arising from asbestosis, and not asbestosis itself, the WCB denied Mrs. Barton a survivor pension.

Mrs. Harold Shorting was widowed in 1978. After 30 years with the company, her husband was awarded a 15 per cent disability pension. Because he died of a complication from asbestosis, the WCB denied her survivor benefits.

Doreen White's husband Fred died of mesothelioma in 1977, after 29 years with the company. In 1973 the ministry of health told him he had asbestosis and should report for annual chest examinations. In January 1977 he was told there was no change in his health. He died in September, on the first day of the week. The WCB deducted four days from his last pension cheque. "I believe that the Health Department gave Johns-Manville a legal license to murder," Doreen White told the commission. "And they did it."

John Dodds couldn't convince the ministry of health he had asbestosis. He said he'd have to die to prove it, his wife told the commission. "We always thought doctors practised the Hippocratic oath. A tragedy had to happen to find out how wrong we were," she told the commission. Our lives "just look like a very bad earthquake. You are surrounded by debris and graves. You are scarred for the rest of your days, and you still can't do a bloody thing about it. Damn it. Do you know what Johns-Manville has done to me?"

Odette Dodds faced the commissioners. "This is my dearest John Dodds, what's left of him. I hold in my hands 40 per cent disability." She quoted her husband's last words to her: "Dying is easy. Going on living and fighting for a cause, human rights, it's the hardest part."

The commission reported in May 1984. The four-year wait for publication allowed the critical issues to die on their own. The Scarborough plant closed in 1980. Asbestos use in Canada was down dramatically. Johns-Manville had made its killing and walked away.

Though it ranked in the top 200 of the Fortune 500 in 1982, the company declared bankruptcy and washed its hands of more than sixteen thousand lawsuits. The WCB was left to pick up the compensation tab for asbestos victims without any company contributions.

The three-volume report made 117 recommendations. It argued that certain forms of asbestos and certain processes for handling it should be banned. It also said most mining and manufacturing processes could be continued under strict controls. That was a "societally-acceptable industrial risk," the commissioners ruled. So was the continued export of asbestos.

The commission also recommended that asbestos insulation in buildings stay put, since the dangers of removing it outweighed the dangers of leaving it alone. It called for reforms to the WCB that would let asbestos victims receive compensation without having to prove beyond doubt that asbestos was the cause of their sickness.

It rejected the major suggestions of unions and Manville employees and survivors. Most unions favoured a total ban on asbestos. The commission disagreed. Leaders of the Asbestos Victims of Ontario wanted criminal negligence suits filed against company, government, and medical officials who denied basic information to employees. The commission disagreed.

Since 1984 workplace safety concerns have been overtaken by campaigns to remove asbestos from public buildings. "That's more cruel than the NIMBY (Not In My Back Yard) syndrome," chief commissioner Stefan Dupre says. "They don't care how big the risk is for construction workers. Because that message never got across, I felt an awful lot of my work went right up the flu."

Manville victims dropped out of the news. Ed Cauchi told a legislative inquiry on October 8, 1985, that 31 per cent of the deceased asbestos workers had been denied any pension. After years of rehabilitation himself, he was told to work as a gas station attendant. "An optimist," he told the politicians, "is a pessimist who lacks experience."

Manville's Scarborough plant also dropped out of the news. The building's new owners, Manson Insulation, kept on some 70 ex-Manville workers, who have stayed with the Energy and Chemical Workers Union. "Manson is a hell of a lot more ruthless than Manville, " local union president Ken Montgomery told Our Times magazine in November 1987. In this plant, silica is widespread, and it's as harmful to the chest as asbestos.

Montgomery says, "Working conditions have worsened. As far as dust is concerned, the working conditions never improved. They are still hazardous." Montgomery is 50, has worked at the plant 26 years, and was diagnosed as asbetic in 1986.

After the United States banned asbestos products in the early 1980s, the Canadian government encouraged exports to the Third World. Canada exported over 700,000 tons in 1987. In 1988 The Economist called asbestos salespeople the "merchants of death." There is widespread evidence that young children in the Third World unload and handle asbestos without any safety precautions. The Steelworkers and the Quebec Federation of Labour support exports. Asbestos provides jobs for Quebec miners, and it can be made safe, they say. Steelworkers booed the Oil, Chemical and Atomic Workers' Tony Mazzochi at a 1982 Montreal conference on asbestos when he demanded that the skull and crossbones be placed on each exported bag. In 1990, as a result of Steelworker pressure, the CLC refused to debate the issue.

Many commission proposals were in line with upcoming reforms. The commission wanted a nation-wide hazard-alert system to mark all dangerous products. That proposal is the kernel of WHMIS — the Workplace Hazardous Materials Information System — in place since 1987. The commission favoured an independent tribunal to hear appeals of WCB decisions. Ontario adopted that in 1985. But the commission rejected a compensation proposal that went to the roots of the Johns-Manville catastrophe.

"I think that the history of the introduction of asbestos into the workplace and the environment is a blueprint for failure," the ECW's health and safety specialist Dan Ublansky told the commission — "failure by industry, failure by government, failure by the medical and scientific community, and perhaps even to some extent of the labour movement itself." He asked commissioners to look at the system, not just its experience with one substance. "The system itself failed to protect both workers and the public from the dangers," he said.

Ublansky asked for emergency relief to victims and their families. They should be treated with the same compassion as victims of any disaster, he said. For the long term, he favoured a new system of workers' compensation modelled on New Zealand's comprehensive disability insurance. He said the only way to deal with low medical knowledge about industrial disease was a no-fault system that gave benefit of doubt to injured workers. Ublansky cited the present uncertainty in medical opinion and said the demand for medical certainty in compensation claims puts the cart before the horse. He asked the commission: Are you "going to hold your victims out there, waiting until all these answers are forthcoming, or are you going to take care of your victims and then pursue the answers?"

This was one of the first union appeals for a no-fault system of compensating industrial disease. The idea is usually associated with Paul Weiler, whose three-volume study of Ontario's WCB cast him as an architect of compensation reform in the 1980s. Ublansky got the idea from Terry Ison, an Osgoode Hall law professor who has championed the reform since the 1960s. On humanitarian grounds, the policy is designed to spare diseased workers from fighting compensation claims on their deathbed. On policy grounds, it applies economic pressure on corporations and governments to speed up research and prevention. It also saves taxpayers from footing the welfare bill when injured workers get turned down by the WCB. On administrative grounds, it eliminates waste on claims and counter-claims. Most of the money kept from diseased workers is pocketed by lawyers and bureaucrats.

The plan has one major problem. By giving the benefit of doubt to the worker, it treats workplace chemicals as innocent until proved guilty. For many people, that's leaving too much undone. Except for prescription drugs and some food additives, thousands of new substances enter the workforce every year without testing. These chemicals are considered safe until proved dangerous. No-fault insurance for workplace disease would quickly force governments to rethink the costs of that approach. That's why Ublansky finished his appeal by calling for a system-wide change in the way workplace chemicals are approved.

"The stakes are simply too high as we head into a new era of microtechnology and biotechnology. Both industry and government begin with the assumption that there is a right to expose people and the environment to potential hazards unless and until such time as a clear and present danger is identified," he said.

Commissioners admitted that the failure to screen workplace chemicals and the slow development of protective regulations cast workers "in the role of human guinea pigs." But the commission didn't call for change.

Another time-bomb like asbestos lurks somewhere, untested, in some workplace in Canada. Only the superficial lessons of Johns-Manville have been learned. "By leaving the employees in the dark about the danger, asbestos, the magic mineral with dust that kills, the biggest industrial killer in history, they were able to abuse the employees, taking away the best years of their lives for their own purpose and fattening their own pockets," Odette Dodds told the commission. "And that is why at this time of age we are having a disaster ... dying for a living."

THE LAST TABOO

Work is a four-letter word that can't be talked about in polite society. Lots of people "talk shop" and complain about their stupid boss and lousy pay. But few people will say how they feel put down and used, how they wish their job could mean something, even though the polls show that's how they really feel. Raising those issues in public seems to be more embarrassing, more intimate, than talking about safe sex, abortion, pornography, or AIDS. Which goes to show that work, not sex, is the last taboo.

Unions operate on the edge of the taboo. They talk about improving working conditions,

by which they mean shorter hours, more safety, less speed-up. And they talk about rights on the job, by which they mean a grievance procedure that allows workers to "obey now, grieve later" if the contract is broken.

The Energy and Chemical Workers have defied this taboo. They say the very way work is organized is beneath human dignity. They bargain for increased industrial democracy, for ways to give workers daily rights to master new challenges, to manage themselves. They're open to changes that workplace democracy will bring to their own internal power structure.

At Westbridge Computer Corporation in Regina, a union-

management program pushed the delete button on the old ways. Career planning decisions are made by both company and union, clerical workers do the hiring for their own sections, programmers work out their own shift schedules, and "hackers" — computer whizkids who don't come on-line for mindless authority — do their own thing. The local union plays a greater part in everyday life, doesn't suffer bouts of indifference between contract talks.

Local president Edie Slugoski says the changeover is hard. "A lot of people develop a real hatred. You want that person to lose. It's hard to give up that hate."

Workplaces like this are still rare and

TAKE THIS JOB AND SHAPE IT

controversial in Canada. It's not just bosses who think such methods are utopian and unworkable. Union leaders do too. Russ Pratt, staff rep for the local, knows the line: that workers let down their guard when they share too many decisions with their bosses, then the bosses do them in. Pratt says that's a smoke-screen, used by union leaders who don't trust their members. He says, "As staff rep, I used to like being the hired gun, the guru. But here, you have to move leadership down to the executive, and get the executive to move it down to the stewards. If you get your kicks out of working with people, rather than for people, you'll try it."

Neil Reimer learned quickly what it meant to be under the gun at work. When he was hired at Regina's Co-op Refinery in 1942, he was told to report for the midnight shift. He'd never worked nights before, so he grabbed a nap after dinner. He slept through the alarm and by the time he got to work the gates were locked. No problem, he'd just crawl under the fence and slip in unnoticed.

He didn't know how tight security was. This was wartime, and refineries were prime targets for industrial sabotage. Reimer says, "The first thing I knew when I looked up, here's a security guard with a gun to my head. Some welcome."

It was enough to give anyone second thoughts about being an employee. It had nothing to do with working hard. It had everything to do with working for someone else. He'd done his share of backbreaking work on his dad's farm, but there at least darkness set the limit on the working day. Personal and social values set the limit on the working week. When he started at the refinery he lost control over his time.

"It caused me to sit back and think," he says. "It was a conscious change in values. On the farm, there was nature and you. But in industry, somebody came in between — the employer, who dictated your values. The concept that the person who owns property has power to dictate a master-servant relationship is the failure of industrial society."

A few years later Reimer got jolted again. The plant superintendent broke from tradition and hired foremen for the first time. That bruised everyone's ego, made the workers feel they'd lost stature, that they weren't trusted. "As a kid of 16 I was in charge of seeding. The livelihood of the family was in my hands," Reimer says. "But nobody sat on the tractor with me to make sure I was doing it right." That also tended to be the atmosphere at the co-op until foremen were brought in. "Nothing happened unless we did it. We had to look after ourselves, and fix our own equipment. All of a sudden, I had to be watched. That's a terrible feeling when people start doubting you. And that feeling has stayed with me ever since."

In 1958, when he proposed that the Canadian Council of the Oil, Chemical and Atomic Workers take the lead in building a new labour party (which became the NDP), Reimer took a swipe at unionists who agreed to "management rights" clauses in collective agreements. These clauses, always at the top of the contract, concede that the owners have the right to do whatever they want with the property and the workers they've paid for. Reimer denounced that as a betrayal of labour principles and urged the new party to take up the issue.

Two decades later, Reimer won acclaim and notoriety as the only national labour leader to push for Quality of Working Life (QWL). The news had just come in from Europe and Japan that workers managed better on their own. Workers there knew their jobs, what the problems were, and how to fix them. New concepts of "participative management" emerged, and managers in North America began to worry that workers might forget their place, insist on rights beyond the installation of a suggestion box.Union leaders fretted that workers might forget their place, start thinking they were too good for a union. They snickered about the "Energy and Comical Workers," who didn't know where the lines were drawn. Reimer never could figure out what the fuss was about. "We were just reinventing the wheel," he says. "We knew it worked because we'd been doing it. The union had to win those things back."

But there's a reason why it took more than 30 years to get the wheel rolling again. Reimer had been slogging it out just to get to first base on the "bread and butter" issues of the workplace: the right to organize, to bargain with bosses as equals, to have better wages and hours. Getting beyond those thorns to deal with the "bread and roses" issues of human dignity at work was hard. There was no support for those issues in either society at large or the industrial relations system.

There was no charter of rights for the workplace, no built-in protections for free speech or equality, no obligation to treat workers with respect, to provide a chance to learn and grow. In the eyes of the law, those rights were bought and paid for. As a result, increases to workers' rights could only be achieved and exercized through unions. But the rights of unions and union members were formal, contractual, tied to the institution — not informal, inbred, tied to the human being. Workers got to vote in their union, not in their workplace. That's why unions were limited to indirect or representational democracy and couldn't make the breakthrough to direct democracy. That's why most experiments in worker participation have collapsed. Direct democracy in the workplace wasn't possible as long as employers gave in grudgingly to union rights, and never to individual rights.

Reimer was stuck in this system as much as any other union leader. But he never forgot where he came from. When a chance came along for him to do something different in the late 1970s, as better managers started to count the costs of old ways of doing things, the union took a gamble on a bread and roses style that would put power on the shop and office floor. The willingness to take that gamble is a hallmark of the union.

It's easy to think the Energy and Chemical Workers' Union was predestined to lead a movement for workplace democracy. Certainly its members had a lot of things going for them.

The values of direct democracy evolved naturally from the experiences of the union's founders, long before prophets from another country developed theories about "participative management" and "Quality of Working Life."

The union's leaders, Neil Reimer and Reg Basken, were both Saskatchewan farmboys. That didn't make them less militant, more militant, less radical, more radical, or less class-conscious, more class-conscious than others. They just had a different sense of workers' problems and rights than leaders raised to accept factory discipline as second nature. Both felt the culture shock, the loss of independence, of the industrial order. They didn't take that particular order for granted, they took it as an insult.

Many of the union's leaders and staff learned their trade dealing with managers of co-ops and crown corporations. The Co-op Refinery, where Reimer became local union president, was part of the farmers' crusade to free themselves from the stranglehold of far-away corporations. Basken was president of the local at Sask Power, which saw itself on an enlightening mission: Farmers and small townspeople rated electric power as high a public achievement as medicare. In Sarnia, Polymer produced synthetics that were key to the war effort and central to Canada's place in a modern industrial economy.

It wasn't sacrilege for a union to take pride in these companies, to help them succeed. It wasn't sacrilege for managers to recognize unions. Both sides could focus their attention on solving problems, not settling grudges.

Union values weren't consistent with the aims of huge, profit-making corporations, but shop-floor and industrial practices in the oil and petrochemical business kept the values in reserve. The union dealt with companies that understood the big picture. Shell, which co-sponsored an experiment in direct workplace democracy with the union in Sarnia, was one example. Shell dealt routinely with forward-thinking unions in Western Europe. It spent big bucks to keep in touch with new theories of industrial management. It could afford to try something new.

Both the energy and petrochemical industry and the union grew up in the post-war boom. They had no point of comparison with the 1930s, when dreams were depressed as much as the economy, when workers were grateful to have any job, when decent wages and job security seemed fantasy enough. The post-war generation of workers expected more from their bosses and unions. Exposing brutal exploitation, ignorant and vicious bosses, didn't get results. The union had to be one step ahead of the bosses, not one mistake behind. It had to offer positive programs that addressed the "higher needs" of workers who were paid and treated fairly well.

As well, the core of the union worked with "process technology." In an assembly operation, the bosses' job is to get those on the line to work harder and faster. It's not for the hired hands to reason why. In a process operation, the bosses' job is to get people to work smarter, so they can move the right dial and push the right button at the right time.

Process technology demands feedback and readings from machines and workers. Haste makes waste if workers just do what they're told, wait for an order before they react to an emergency. Progressive management methods go with the flow of process technology.

Despite this unique social and industrial background, the OCAW did nothing to promote direct industrial democracy experiments throughout the 1950s and 1960s. There was no foundation for it when the union had to fight hammer and tong for basic recognition and survival. Every difference was a test, every concession a defeat. So conflict over basics bred conflict over slights: Give no quarter, ask no quarter. Even at Sask Power, where collective bargaining was freely accepted, union leaders were leery of any moves that might lead workers to let their guard down. The pro-labour CCF government in Saskatchewan offered the union seats on the corporation's board of directors, according to Thomas and Ian McLeod's biography of Tommy Douglas. But union business agent Cy Palmer turned it down, arguing that it would "soften up our people too much." The union also felt that having members on the board would create a conflict of interest situation.

In 1963 Reimer had labour-management co-operation problems of his own after the International president demoted him for bucking a presidential decision. Later, when he returned as union district director in 1967, he was full of fresh ideas and broad perspectives. The touring he had done in the interim as Alberta NDP leader had introduced him to the widespread poverty that unions neglected. In his farewell speech to the NDP he urged activists to press for industrial democracy as a way to give people a sense of belonging and purpose in an anonymous society. He no longer had the time or patience for internal union squabbles and small-change union politics. He wanted to deal with issues that would renew the union's reason for being. "I knew that the union had to mean more than nickels and dimes. We had to press new issues," he says.

Despite this personal commitment, it took ten years before the union could move ahead on the new issue of industrial democracy.

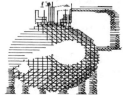

The most far-reaching union venture into the frontier of direct workplace democracy took hold at the Shell chemical plant in Sarnia, Ontario. The plant, set up right next to the Shell refinery, is a showcase of futuristic technology, efficiency, and work organization. The $200 million dollar operation employs two hundred people who turn petroleum feedstocks into the makings of antifreeze, rubbing alcohol, solvents, rope, carpets, and plastic fixtures. Those are tough and exacting products, and they have to meet the price and quality standards of a world market.

Established in 1979 after five years of planning, the plant went on to break every production record in the business. It also broke every rule of traditional management. The entire operation still runs, 24 hours a day, seven days a week, with a bare minimum of supervision. Workers and co-ordinators have divided themselves into six operating, or "process," teams and one maintenance, or "craft," team. Each team manages its own affairs.

The social set-up of the plant wows efficiency experts from around the world. It's considered one of the most far-ranging experiments in worker self-management anywhere. Everything about it is new. It's a "greenfield site" that broke ground in design as it broke ground in construction. Unions and managers worked as equals in the design plans, training programs, and hiring decisions. It has the shortest collective agreement in the country, but managers and workers negotiate 365 days a year over everyday details of plant operations.

The plant's inner workings have been put under the microscope of two Ph.D. dissertations. Norm Halpern, formerly a senior Shell manager, wrote up his insider's account for the Ontario Institute for Studies in Education. Tom Rankin, a consultant who worked closely

241

with the union, wrote his for the University of Pennsylvania. Rankin calls the union leadership "visionary" and hopes the experiment represents "a probable future in the present."

Canadians, predictably, have barely noticed the plant, except to comment on its success in reducing labour-management conflict. In fact, the plant's success is the product of passionate, intense, and ongoing conflict. Outsiders are misled because the conflict is mature and creative. That's because both sides were up-front about their needs, didn't cloud the agenda with claptrap about peace and togetherness, and got together to "work through" their differences.

Early in the 1970s, Shell executives became aware that labour problems stood in the way of corporate expansion. Company refineries couldn't hold onto experienced workers, who left in droves to escape the grind of constant shiftwork. The company faced a chronic shortage of skilled maintenance workers, who could mix knowledge from a few construction crafts to meet less specialized industrial needs. A good two years before the chemical plant got to the drawing boards, top management was open to trying something different and people-oriented.

In 1973 a senior management team drew up a checklist of human resource needs in tune with the latest wisdom in progressive management. The report was hardheaded, put the company's personnel needs first, and didn't delude anyone with tirades against overpaid workers and tyrannical unions. A good five years ahead of the competition, Shell realized that a carrot makes better chemistry than a stick.

High wages had to be offset by higher productivity, faster decisions, and less staff turnover, the report said. The company had to rely on self-starters, people who thrive on challenge, variety, and flexibility, who don't want to be hassled by senseless rules and authorities. To attract such workers, the workplace would have to be more open and democratic. The report got the nod of the manufacturing vice-president, who summoned Norm Halpern in 1974 to lead the way towards organizational efficiency.

A 1975 plan for the chemical plant bears the stamp of the most avant-garde Quality of Working Life ideas from the Tavistock Institute in England. The 1975 report said technology must be adjusted to people, not the other way around, to get the quick response time and quality decisions that make technology hum. Those qualities couldn't be bought. They had to be earned. The new plant must offer meaningful and challenging jobs, a "climate which encourages initiative, experimentation" even when mistakes are made, and a minimum of status differences.

The design plan was formally adopted in October 1975, and go-ahead was given to seek union approval. Reimer was keen to try but insisted that the union be treated as a full partner and get a high profile. Shell agreed and appointed the union president of the Sarnia oil refinery to its design team.

The designer-plant broke all the rules of corporate architecture. Drafts completed by British engineers, drawn with people as an add-on to technology, were junked. The standard separation of scheduling, maintenance, and quality-control departments was canned. Special offices for the status hierarchy and special perks like separate parking spaces and cafeterias were dropped. Instead the plant was built to house self-managing teams with all the skills needed to handle every basic job. The teams worked out shift and vacation schedules among themselves, and they worked up their own plans for training. Two levels of unnecessary supervisors got dumped.

Inventors were told to make a machine to pack bags. The work was considered too boring and dirty for humans. The smart machines lost some of their innards so that people could make the artistic decisions in the creation of polypropylene, plastic pellets. Computers were rewired to give nudges and hints, not orders. "Try for 350 degrees; it's worth a saving of $550," a typical read-out suggested. The big offices went to people with bulky workloads, not bulky titles. Punch-clocks were trashed. Workers got paid for what they knew, not the particular job they did. All workers had access to training programs that would enable them to earn higher pay.

Everything seemed to speed along without a hitch. Management was on side, and union leaders were on side. Everything was going to be state of the art. Then Shell refinery workers, who were neighbours and potential recruits to the new chemical plant, were offered a chance to opt into the dream workplace. In October 1975 the executive put the issue to the refinery workers, without making a recommendation. By a vote of 112 to 3, the workers turned thumbs down.

The theorists had overlooked some essentials of Quality of Working Life from a worker's standpoint. While an oil refinery is clean, a chemical plant is dirty, smelly, and hot -- a negative that desegregated parking lots can't counter. The chemical plant had an ingenious plan to keep night-work to a minimum but it still operated on shifts. Some of the refinery workers had built up enough seniority to work straight days. Others had purposely taken demotions to get day jobs. They weren't prepared to give that up to take part in an experiment.

The set-back strengthened the union's hand. Because the refinery and chemical plant were next door to each other, and because they were both owned by Shell, the union was in a strong legal position to insist on covering both plants with one collective agreement and one seniority list. The staff rep for the Sarnia area, Stu Sullivan, told the company that if it wanted a fresh start, with a new workforce and a separate collective agreement, "You have to help us, recognize our needs." He said, "If we're going to take a risk, we need union security."

Sullivan convinced Shell to grant voluntary union recognition and compulsory payment of union dues, the so-called Rand formula, in the new plant. That formula, standard in most unionized industries, had always been rejected out of hand by oil companies.

Sullivan was no slouch when it came to talking up new management theories. He had taken a crash course on democratic plant design with Lou Davis, a world leader in the QWL field, and had co-authored an article with him. Sullivan explained the importance of working democratically to achieve a flexible workforce and a participative workplace. Input from workers had to come early and often, he said. If workers only had a choice between accepting or rejecting a finished proposal, they'd likely see the proposal as a company ploy, a shell game, and vote it down.

The set-back put a fine point on Sullivan's warnings, and management took note. Through 1976-77, union and management worked on technical details and a plant philosophy based on the principles of value-driven socio-technical design — in plain English, putting people in the driver's seat of technology.

By the time the plant had opened its doors in 1978, union and management interviewers had screened some 2,600 applicants to pick the right mix of social and technical skills to fill a maximum of 200 jobs. A Company of Young Canadians volunteer, an ordained minister, a horse-trainer, and an insurance adjuster were on the founding union executive.

Judy McKibbon was one of five women hired in this traditionally male industry and one of two women on the first union executive. McKibbon, an active feminist, says, "We have already made decisions to take control of our lives. So it's natural for us to move into self-management at work."

McKibbon was amused by the fuss about the plant's unusual set-up, referred to on the inside as "this socio-tech thing." As she sees it, "If we can make decisions at work, surely we can make decisions about work. Our livelihood is too important to leave to managers. If we leave it to them, all they do is stand over you with a whip and say do it faster. But there's a way of doing it better too, and we have something to say about that."

A long-standing chief steward for the local, McKibbon has the job of breaking in managers who regard union consultation as a problem-solving gimmick. She tells them that it's more than that. "It's a value system," she says. "It's a way to expand collective bargaining infinitely, to move in the direction of self-management."

The company's investment in this new breed of workers paid off. Corporate audits in the early 1980s rated productivity at 175 per cent above original specifications. The productivity target for 1990 was 250 per cent above. Quality control is rated excellent. Repair bills are kept low because of regular checkups by multi-skilled work teams. "With technology that costs $50,000 a minute to shut down, we can save a lot of money by plugging the dykes," McKibbon says.

Norm Halpern's thesis rates union-management relations as excellent. He counted 11 grievances over four years, compared to 87 in Shell's refinery next door, which has about the same number of employees. That count is deceiving if it suggests that the relationship has been trouble-free. The chemical plant avoids settling disputes through grievances — formal complaints about contract violations that go to an outside arbitrator if they can't be resolved. The contract weighs in at 13 pages, a good 70 pages shorter than most. Until the late 1980s it made no reference to formal grievance procedures.

Instead of standard dry-as-dust legal phrases, the collective agreement leads off with a promise to "ensure an efficient and competitive world-scale chemical plant operation and provide meaningful work and job satisfaction for employees." The preamble vouches that "employees are responsible and trustworthy, capable of working together effectively and making proper decisions," and says it's necessary "that a climate exists which will encourage initiative, experimentation, and generation of new ideas, supported by an open and meaningful two-way communication system."

The contract has not been tested by an arbitration board — which makes decisions on the basis of tight wordings, not fine phrases — to see if it's worth the paper it's written on . So far, there hasn't been a need. Conflicts have been settled directly and, when necessary, by reference to a "good practices handbook" — an open handbook, written in plain English and adjustable to fit new circumstances. "We negotiate 365 days a year," one unionist says.

Unionists blame high management turnover, especially among engineers, for most problems since the mid-1980s. "They're on their way up, or on their way down, when they go through Sarnia," Sullivan says, "but that's what they do. They go through Sarnia." As a result, he says, the union has been left to maintain continuity: "It's become a union program." Managers may accept decentralized decision-making but they haven't bought into shared power. "The workers accepted more responsibility, but they insisted on getting authority along with it," Reimer says. "That's really what the core is."

The program started to drift badly in the late 1980s, McKibbon says. "Once we got through the razzle-dazzle issues and into the subtle ones like discipline, refresher training, and technological change, things moved too slowly." By 1988, she says, "The union felt it was on shaky ground, with no way to protect people, so we insisted on a clause that recognized the good practices handbook and philosophy statement could be taken to an arbitrator." With 12 hours to go before a strike deadline, the company gave in.

Managers aren't the only ones with problems accepting "this socio-tech thing" as a package deal. Union leaders and structures also had to adjust to a local that involved 90 per cent of its members in committees. "Our members felt they had a hands-on policy with their jobs and they should have a hands-on policy with their own union,"

Reimer says. "There isn't much point to owning a car and not being able to drive it."

Reimer has had a soft spot for the local since he watched it put on a steward-training program in 1981. He could feel the difference in levels of confidence, authority, and responsibility, he says. "It was a tremendously impressive display in individual development." Sullivan left the meeting thinking, "We can't change the organization of work without changing the nature of the union." He adds, "If workers have authority lower down in the corporate structure, the union has to to push it down too."

The steward has to become the building bloc of the union executive and has to be trained in problem-solving, not grievance-filing. Union meetings are needed for accountability, Sullivan says, but they have to be accepted for what they are — boring. Active involvement has to come through other routes.

The Sarnia experiment came in the midst of a stormy period in the union, when there were big question marks around the role of U.S. labour connections and around pollution caused by the energy and chemical industry. In 1977 the Canadian Council of the Oil, Chemical and Atomic Workers held a special conference on QWL. Reimer tried to tie the choices together. A new, independent union has to consider "the type of decisions that you may wish to make for yourself in the plant, and the recognition that we're going into a conserver society," he said. That called for "a total review of what really are the rights of workers."

Union shakeups had the same impact on the mutineer staff who formed the Canadian Chemical Workers. Russ Pratt and Bill Adams serviced 750 members at the giant fertilizer plant, Cyanamid, in Welland. In their 1979 bargaining they learned that the plant was headed for closure, and they called on the company to involve the workers in rescuing it. "As a brand new union, we had to make a difference, and had this belief in the members and their importance," Pratt says. "We have such little impact on wages, we have to do something more relevant, help workers become whole people."

The plant had nowhere to go but up. The waste in the plant was awful. "The place was filthy," Pratt says. "We were expected to do our jobs like robots, to put up with foremen breathing down our necks, and to shut up," plant worker Bob Roberts said. "No one believed we had an idea worth listening to. If you weren't an engineer with some kind of degree you didn't count," he told consultant Sheila Arnopoulos, author of *Particichange: A Quality of Working Life Program in a Chemical Plant.*

Bill Adams says the union was a part of the problem. "No one took the time to see what the problem was, so we went after more money. For the company, that was just the cost of doing business," Adams says. At the first training session for the new program, the company picked up the tab for the bar. Pratt and local union president Al Wood saw two unionists carting off boxes of booze. "If they didn't want us to take it, they would've put a guard there," the two workers explained. "That's where we came from: beat them anyway you can," Pratt says. Pratt and Wood smuggled the booze back.

In 1983 the company credited the program with savings of $372,000, against costs of $158,000. But it wasn't enough to overcome high feedstock prices from the National Energy Program. The plant closed in 1983.

In the 1990s unions will have to adapt to the new realities of a high-tech workforce, Reimer says. "As computerization increases, and workers on the shop floor make decisions that middle management made ten years ago, how can the union be immune from that process? If we can't deal with it internally, how are we going to attract people to join us?"

Then Reimer reverts to type, the farmboy who doesn't like being bossed around, the old-style union crusader of the CIO days who asked workers to stand up and be counted. "That's my view of what the union's about. When we ask workers to stand up, we mean on their feet not ours," he says.

"What's the difference with being dependent on a company? The union's job is to help them become whole persons in an industrial society."

Bringing it all back home

TWENTY QUESTIONS

Americans think we're pretty much the same as them, huh? Except that we ask more questions, eh? Questions about health care being for sale, about melting pots where racism, crime, and bitterness stew, about a land of opportunity where millions sleep on the streets, about a democracy where only half the voters bother to cast a presidential ballot, about basic public services being run for profit, about one country's right to interfere in another.

When the differences are boiled down, many of Canada's different questions and answers came from the labour movement and alliances it built. The Canadian identity was created, not handed down.

Canadian unions aren't just U.S. unions with snow on them. Unions that represent more people, win better contracts, and have a greater say in public affairs than south of the border took leaders who understood the difference between running a union and organizing a labour movement — who understood that unions were not just an interest group and an opposition, but a public interest group

with an alternative. Some explanations go back hundreds of years. But the strategic difference came in the 1950s, when Canadian unions moved to adopt their own political party.

U.S. unions bowed to pressures to limit their programs to what could be achieved in a two-party system. Canadian unions helped to build a party to express their own ideas. That has made all the difference, not just in the formation of a viable political opposition, but also in the range of community activities and policy options to which Canadian

EVERY SQUARE PEG

unions subsequently contributed.

Reimer had both back-room and front-row seats on these changes. In the 1950s, the Co-op Refinery and Sask Power locals of the Oil Workers International Union were in the thick of Saskatchewan's test drive of social democracy. In the 1960s, oil and petrochemical locals in OCAW were decisive in Alberta, where Reimer led the NDP. In the 1970s and 1980s, the energy and chemical complex was centre-stage in raging debates on the environment, Native rights and Canadian industrial strategy.

Canada's wheat belt in the 1940s was the last place anyone expected to find basic bread and butter unionism.

In 1944 Saskatchewan workers and farmers elected the first socialist government in North America. The CCF government made the province into the place to be to explore the wide open spaces of social change and ask the really big questions. Saskatchewan became a laboratory for social and political experiments of every kind, from adult education to public auto insurance — the perfect place for workers and their unions to establish a showcase for labour reform.

To unionists in the rest of the country, Saskatchewan looked like paradise. The grass didn't look so green to Saskatchewan unionists. The CCF was a grassroots movement, to be sure, but it was primarily a movement of farmers, who tended not to understand workers' problems. For that reason, Gad Horowitz writes in *Canadian Labour In Politics*, Saskatchewan union relations with the CCF were "unique." Labour's "most difficult task," Horowitz wrote, "was not to win the unionists'

251

vote for the CCF but to strengthen union influence within the CCF in order to counteract anti-labour trends in its farmer majority and in the CCF government."

Oil and power workers were part of that unique experience. For that reason, they took the lead in defining union relations with public-sector and other employers. They also led the way to the formation of a new, more labour-based, party that replaced the CCF, the New Democratic Party.

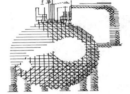

Neil Reimer started working at Regina's Co-op Refinery when the big questions were subjects of constant debate during work breaks and union meetings. "Because the movement was so politicized and because the role of co-ops was so important," he says, "we had to decide what role the union movement would play. The role of co-ops, socialism, and democracy were all part of it."

Co-ops were at the centre of Saskatchewan's political and social ferment. The CCF, after all, was named after them. Its vision of the future was a co-operative commonwealth federation, not the faceless socialist state its cold war opponents slandered. Far from relying on the state, Saskatchewan homesteaders formed co-ops because the state was too far removed from them, both geographically and politically. Co-ops were closer to home, like quilting bees and neighbourhood barn-raisings.

The Co-op Refinery was the crown jewel of the co-op movement, offering an alternative to the world's most notorious monopoly, the only such outfit in the world. It was a symbol of the co-op political and social mission: an alternative to competition among farmers, to domination by Eastern monopolies, and, many hoped, to labour-management conflict in the workplace.

During the war, the Regina Co-op's union was torn by conflicts between CCF and Communist Party supporters. Reimer cast a plague on both houses. He couldn't see splitting the difference between two visions of a far-off future, when workers across the street at the non-union Imperial Oil Refinery made better money in the here and now. "Let's concentrate on that before we get into the fancy talk," he said. He ran up the middle and won the local's presidency against both factions.

His hardline bargaining stance reflected his views on co-op-union-government relations in an agrarian province, views that did not include a preference for simple business unionism. "The attitude in the co-ops and CCF was `you don't need unions; you have us.' I always thought we just needed laws so we could do our job, not so they could do it for us," Reimer says. That sense of self-reliance held true to the farm traditions of both the co-op and CCF movements.

Precisely because he was so steeped in these agrarian values, he understood where they didn't apply in industry. "Farmers own property, so they don't understand what it is to be a worker and have somebody stand between them and their work," he says. In his view, workers needed unions to underpin their independence just as farmers needed co-ops to underpin their's.

This commitment to social independence and hard-edged bargaining gave his union views a gritty realism that didn't sit easily with CCF leaders.

When Reimer recruited Saskatchewan power workers to the Oil Workers in 1954, the government and union got their first crack at a rounded and mature relationship.

The CCF wanted to do right by labour. CCF leaders believed the union cause was just, and they believed that union votes were key to CCF victories on the horizon for the rest of the country. They "had their eye on Ontario," says Ken Bryden, an expert in labour legislation recruited to Saskatchewan's ministry of labour. "We thought the way to make the breakthrough in Ontario was to become a model in terms of labour relations in an agrarian province."

Bryden returned to his native Saskatchewan with a team of about 30 of the best and brightest — "carpetbaggers" they were called by jealous supporters of the old order — determined to give the new government the latest tips on good government. They were a crack team, headed by George Cadbury, who renounced his family fortune in the chocolate business to offer his business skills to the socialist movement. They gave Saskatchewan the best labour laws on the continent, the highest minimum wage, the first paid holidays, the first 44-hour week, the first Human Rights Code. They funded co-ops, adult education, and hospitalization. They set up public fur marketing to help Native trappers, expropriated a box factory to penalize an anti-labour employer, and nationalized hydro to take electricity to isolated farmers.

But there were limits to what even a sympathetic and energetic government could do. With the best will in the world, the CCF had to win elections in an agrarian province and run a government on the brink of bankruptcy. The CCF was also a prisoner of its time. Its human rights code, for instance, made no reference to women. Its labour act made no reference to industrial democracy, a step that "would've been very difficult," according to Cadbury.

"Farmers are not pro-labour, and they would've seen us as handing over control to the unions. That's not an adequate reason not to experiment," Cadbury says. "We could have done it, and we should have done it. It was an opportunity missed."

253

Premier Tommy Douglas had a favourite story to illustrate the farmer-labour relationship, about a farmer about to have a painful operation. He'd heard about a new painkiller called Twilight Sleep and asked if he could try it. But Twilight Sleep was for women giving birth, so the doctor turned the farmer down. "It's just for labour," the doctor said. "That's the problem with this damn province," the farmer screamed, "everything for labour, and nothing for us farmers."

Bryden left Saskatchewan in 1949, knowing the province had gone as far as it could. "It wouldn't have been wise to do more," he says. "They were already so far ahead. People still hadn't digested what we'd rammed down their throats." Cadbury left in 1950, when he sensed the glory days of change were over.

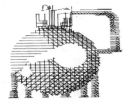

When Reimer asked Sask Power workers to join the fledgling Oil Workers in 1952, he made no bones about his willingness to take the government on. That's the main reason why they turned in their charter with the powerful Canadian Congress of Labour to join his tiny union. The CCL, with its strong political commitment to the CCF and fear of embarrassing a CCF government, was seen as a company union. It didn't matter that oil and hydro belonged to different union jurisdictions. It mattered that Oil Workers saw the different jurisdiction between unions and union-backed governments. The switch to the Oil Workers was a mandate for hard bargaining, and hard bargaining was what they got in 1954 and 1955.

They were bad years for farmers. The worst drought and rust in 20 years left most farm families $2,500 short of the previous year's earnings. Premier Douglas banned wage increases for government workers until times got better. "We cannot give what we haven't got," he said. "Under present economic conditions, no group in the community is entitled to a larger slice of a smaller cake."

But his government was trying to have its cake and eat it too, at the expense of power workers. The Saskatchewan Power Corporation was formed in 1949 to take power to the people. It was designed to replace a score of private and city-owned facilities with one system that spread out the costs of serving isolated areas. From 1953 to 1955, when bargaining with power workers became a high-wire act, electricity was introduced to 19,500 rural homes and 76 communities.

According to Clinton White's *Power For A Province: A History of Saskatchewan Power*, financing was complicated when towns refused to give up their own power stations. Many towns made money from hydro sales and used the revenue to avoid unpopular tax hikes. So Sask Power was stuck selling power to big centres at wholesale rates and absorbing the entire cost of taking power to remote areas.

Douglas treaded gently with the power-towns. "We felt that it was better to proceed by negotiation rather than by compulsion," he said. "We were constantly being accused of using `compulsion.' As a matter of fact, we sought to reach desirable social objectives by a process of persuasion wherever possible." But when persuasion didn't convince power workers to cover Sask Power losses with lower wages, he turned to compulsion with them. A classic case of a politician following the course of least resistance.

Bargaining was short circuited by the refusal of managers to discuss anything but a wage freeze. The stalemate at the bargaining table dragged on for nine months. Douglas took his bureaucrats' word that union negotiators had irresponsible demands. He even charged that the union favoured privatization of Hydro if that was the only way to get more wages. A union vote showed 88.8 per cent support for a strike.

It became a cat and mouse game between Reimer and Douglas. Reimer stalled for time. He tried to convince Douglas, "As long as we're talking, even if it's just to move a comma around, we're making progress." Reimer's game-plan was to wait for the legislature to dissolve, so Douglas couldn't counter a strike threat with a back-to-work order. Douglas didn't fall for that. He cooked up any excuse to keep the legislature in session. "Any visitor, any curling team, got invited as a guest of the legislature," Reimer says.

The showdown was coming. Blackwire fever set in, and the party lines rang off the hook. For the labour movement across the country, this was no ordinary strike in an ordinary province. It threatened civil war between unions and their party. Labour leaders were queasy.

"Everybody thought we were too militant," Reimer says. "We acted like a union with a CCF government, and at that time it was hard for people to accept. No one could understand why we were having trouble with those wonderful people in Saskatchewan. I said they better come and see for themselves. I got a committee of the labour congress to meet with cabinet. I knew they had to experience it for themselves. We left that meeting together. As we drove across the Wascana bridge, [Canadian Congress of Labour president] A.R. Mosher said, `This is no longer a CCF government.' He couldn't believe it."

On a Sunday morning, Douglas called the union bargaining team to his office, told them he couldn't afford any more delays in getting a settlement, then left the room, leaving a wad of paper in full view. The unionists took a look at the back-to-work law he'd prepared. Go ahead and try it, Reimer said when Douglas returned, and if you want to throw us in jail, hop to it.

"I spent the whole afternoon worrying," Reimer says. Then a plan dawned on him. First thing the next morning, he called Charlie Millard, leader of the Steelworkers and CCF candidate in an upcoming Ontario election. "I thought you'd like to know that we're probably going to be on strike and the government's going to force us back to work," he told Millard.

Millard was frantic. He was sure this would destroy the CCF's credibility in Ontario. It would feed the propaganda mill with gleeful editorials about the CCF crackdown on labour. Millard put his troubleshooter, Bill Mahoney, on the case. Reimer suggested to Mahoney that David Lewis, prominent labour lawyer and senior CCF official, be appointed mediator. Reimer outlined the settlement Mahoney should pass on to Lewis. "How am I supposed to do that?" asked Mahoney, a man not known for his innocence. "He's your lawyer," Reimer said. "Instruct him."

Douglas and Reimer booked a conference call with Lewis and requested his services as a mediator. "You're the national president of the CCF, so you can't possibly do the CCF in," Douglas told Lewis. "And as a labour lawyer, you can't possibly do the union in. So we think you're the right guy."

In his autobiography Lewis wrote that "No other incident produced the anxiety and drama" of that settlement. Lewis sympathized with the government's need to cut costs and shared its hostility to the union's chief bargainer, Cy Palmer. In Lewis's eyes, Palmer was "paranoid" about government intentions to skin the workers and out to wring every concession from a sympathetic and vulnerable employer. Lewis didn't know that Palmer was Reimer's stalking horse, the straight man who made Reimer look reasonable. Lewis brought the parties to a settlement within ten days. He wrote: "I was able to get away with saying to labour, `stop being bloody fools' and I was able to tell management `you better go back and tell the government this isn't acceptable.' It worked out."

As soon as the crisis passed, Reimer met with Douglas and his finance department head, Carl Edy (Reimer's former high-school teacher) to discuss future bargaining. The government should keep its nose out of crown corporation bargaining, Reimer said. It should hire good managers and make them take the heat. Otherwise, a vicious circle clicked in, where workers pestered cabinet about wages and managers passed on the tough decisions to cabinet. Douglas didn't interfere in Sask Power bargaining again. Reimer says the brush with compulsory arbitration gave Canadian labour leaders their first inkling that there was trouble in paradise, that unions needed a party and a relationship with a party that would allow them to be more assertive. "The CCF did a lot for labour, but we really have to do things for ourselves, and they didn't allow for that kind of involvement," he says. "We could endorse them, but that was pretty passive."

Long before the 1958 federal election, which almost wiped out the CCF and forced labour activists to push for a new third party, high-level union leaders began thinking on new political lines. In 1954, Oil Worker staff rep Roy Jamha told a conference of the Canadian Congress of Labour that it was time for unions to build their own party, not just endorse the CCF. "He was shot down in flames," Reimer says.

In 1955 Jamha made the same pitch again. Donald MacDonald, CCL secretary-treasurer, said the proposal was worth a serious look. In

1956 the Canadian Labour Congress merged the old CCL, which favoured the CCF, with the old Trades and Labour Congress, which favoured political neutrality. The new CLC raised the need for a new labour politics.

"NATION IN A FLAP OVER NEW PARTY" blared the banner headline of the *Winnipeg Free Press*. On the day before, April 23, 1958, the Winnipeg convention of the Canadian Labour Congress had called for "an effective alternative political force based on the needs of workers, farmers and similar groups, financed and controlled by the people and their organizations." Over 1,700 delegates, representing 1,150,000 unionized workers, broke into cheers, chants, and songs. The vote was almost unanimous. Everyone knew that political history was being made. "The time has come for a fundamental realignment of political forces in Canada," the resolution said.

Indeed, a fundamental realignment had taken place less than a month earlier. On March 31, John Diefenbaker's Progressive Conservatives trampled Liberals and CCFers in one of the great election upsets of the century. The CCF was decimated. Party leader M.J. Coldwell was swept aside. So was Stanley Knowles, the champion of pension reform from the safe labour riding often known as "Winnipeg-off-centre." So were 15 other caucus members.

The Liberals, long used to seeing themselves as the government party, were reduced to a corporal's guard. It wasn't clear if they could survive in opposition. Like everyone else they sensed the political vacuum in Canada. When they heard of labour's call for a new party, the *Free Press* reported, they issued their own call: "The Liberal Party must now begin a long painful rebuilding process."

The creation of a political vacuum, and the rush to fill it, cleared the air of Canadian politics. But it didn't clear the record of myths. Labour people, ever the underdogs, reach instinctively for the old Greek classic about the phoenix rising from the ashes. One of the famous lines from the union anthem Solidarity Forever sings of building the new world on the ashes of the old -- the next best thing to a virgin birth. But the vacuum theory of the NDP's birth leaves out too many messy details.

The 1958 election created a bandwagon effect for a new party, but the old party's days had been numbered for some time. The numbers could be seen most clearly in union affiliations. According to Ivan Avakumovich's *Socialism in Canada: A Study of the CCF-NDP in Federal and Provincial Politics*, there were only 14,417 affiliated unionists in 1956. Half of them came from Nova Scotia mines.

At the founding convention of the CLC in 1956, industrial unionists, who had endorsed the CCF as the political arm of labour in

1942, kept a discreet silence about party politics: no use scaring off new friends with old political differences. More to the point, there was no use beating a dead horse. David Lewis, leading CCF strategist, realized that Canada's more mature labour movement couldn't be content endorsing a party; it wanted to be a partner, not an appendage. The Saskatchewan experience of power workers was no doubt a factor in that maturity.

Lewis and other CCF supporters were content that the founding convention of the CLC simply called for political education and exploration of a new political coalition. The politics and players behind that new coalition weren't spelled out. There were hints, however. In 1956 the CCF renounced its ringing declaration for socialism and supported the idea of a "mixed economy." The proposed new coalition had the ring of a mixed party, with no clear stance on behalf of labour.

OCAW led the calls for a new political party and to bring key political issues to the fore. In a bid to force the hand of the 1958 CLC convention, OCAW held its Canadian Council in Winnipeg on the eve of the expected debate on political action. On that Sunday a unanimous resolution called on the CLC "to initiate immediate action" to establish a new party that would express the social, economic, and philosophical aims of the labour movement.

According to reporter Dudley Magnus in an October 16, 1965, retrospective in the *Winnipeg Free Press*, OCAW delegates then leaked their resolution, and hints of a possible convention showdown, to the Winnipeg paper. When CLC convention delegates took their seats the next morning, they read the news that OCAW had jumped the gun and set the terms for debate.

Old-style craft unionists weren't pleased. "Verbal war has literally broken out in Winnipeg between two opposing union officials on the issue of a labour party being formed," the *Winnipeg Tribune* reported. OCAW delegate Alex McAuslane clashed with Jack Sharkie of the International Brotherhood of Electrical Workers. It didn't help that the two unions were competing to organize Manitoba hydro workers. "Oil Chemical is only two years old and immature statements on political action as this one can be expected," said Sharkie. McAuslane replied in kind.

After the lines were drawn on the first day, the CLC executive presented its version of OCAW's resolution. The formal vote on Wednesday was a foregone conclusion. It's a common trick in the labour movement to bury differences in abstract lines that outsiders can't read between. The CLC resolution imagined that the new party would rush into the vacuum created by the seeming destruction of the Liberals and establish a new political realignment somewhere between a pure conservative party and a broad liberal or social democratic party.

OCAW seemed to be the only union to project the new political coalition as a labour party. The union's weekend resolution made that clear, as did the name for the new party suggested by OCAW delegates.

While others were dreaming up names like the New Liberal Party, a poll of OCAW members suggested Labour Party of Canada or Canadian Labour Party.

The Liberals proved that reports of their death were greatly exaggerated. They got the hint from the 1958 election. Under the leadership of Keith Davies, Tom Kent, and Walter Gordon, they set out to create a reform image for themselves in the 1960s. The old Liberal czar C.D. Howe complained that a Liberal think-in in 1960 went too far to the left. "Stop sniping at me," Pearson replied, according to Peter Newman's *Distemper Of Our Times*. "This is practical politics. I don't intend to let the New Democrats steal the popular ground of the left." That tack brought them back to power for another 20 years.

OCAW activists beat the drums for the new party, but the union's reputation was stained by rumour-mongering about Reimer's support for Hazen Argue in the contest for leadership of the new party. Tommy Douglas, a proven winner and moderate, was the odds-on favourite of the CCF and union leadership. Hazen Argue was the outsider, leery of union domination and friendly with left-wingers who wanted the new party to oppose Canadian membership in NATO. Reimer, an old school chum of Argue's, was considered off-side.

According to one prominent party father, quoted anonymously in Terrence Morley's *Secular Socialists*, "Hazen seems to have been able to rally every square peg, both in and around our movement, to his side — the Colin Camerons, the Dorothy Steeves, the Leo Nimsicks, the Cedric Coxs, the Pawleys from Winnipeg, the Bill Seftons and Jack McVeys from Toronto, the Neil Reimers from the Oil Workers, the Doug Fishers and the Arnold Peters." Reimer was mistaken for an Argue supporter, simply because he crossed David Lewis and nominated Argue for temporary CCF leader, after former leader Maj. Coldwell lost his seat in the 1958 landslide. Lewis didn't want Argue to carry any prestige as a CCF leader into the NDP founding convention. Reimer didn't want the CCF to remain leaderless for three years over such petty and purely partisan considerations.

The whisper campaign against Reimer served to discredit his two points of opposition to the proposed profile and structure of the new party. He wanted the new party to declare itself a labour party. Most thought this would saddle the new party with a grimy and militant image, pose class conflict too sharply, and cut the party off from farmers and the middle class, who were expected to realign politically once the Liberal Party disappeared. Reimer believed the party had to reclaim the dignity of "labour," that it had to use the name as a talking point to explain the politics and vision of a new society "that offered a little more horizon than capitalism with some social justice."

Reimer worried that the style of union affiliation to the new party was too influenced by the British Labour Party, which grew out of a more political and class-conscious culture that didn't put as much of a premium as Canadians did on politics as a private affair. He believed

259

that unions should not affiliate to the new party until a majority of members signed up on their own. He considered affiliation without that support to be premature proxy radicalism, a substitute for serious political work, and "a bit of an ego trip to think that workers will do as we say." He also thought affiliation promoted a bureaucratic relationship between union and party leaders, at the expense of membership involvement.

He promoted both of these positions within OCAW and the CLC, where he chaired the political education committee. He remained a lone voice on both issues.

THE OTHER SIDE OF THE STREET

Unions got involved in politics for the same reason the chicken crossed the street — to get to the other side, where labour laws, health and safety laws, inflation, interest rates, full unemployment, and medicare — all of which affect union bargaining — are decided. But when unions cross the road, they're charged with jaywalking.

The best-known case was brought by Merv Lavigne and the ultra-conservative National Citizens' Coalition in 1985. They accused the Ontario Public Service Employees Union of giving members' dues to political causes they didn't believe in, of violating individual rights.

Though the Charter of Rights

changed the ground-rules for these sorts of cases, the arguments aren't much different from those used against OCAW for its support of the NDP in 1961. Shortly before the NDP was formed, B.C. premier W.A.C. Bennett changed the province's labour laws to make it illegal for unions to make political donations. He said he wanted to protect the rights of individuals who didn't want their dues money spent that way. Imperial Oil had held back over $40,000 in union dues from the OCAW to prevent the union from donating a paltry $87.89 to the NDP. When the union sued for the dues owed, it became the test case for union political rights.

The courts backed the government. B.C.'s Justice Whittaker said political parties were the place for citizens to express their political views. It didn't matter if a majority of union members backed one political party, he said, because that had to be an individual decision. The Supreme Court agreed. Unions can force people to bargain together, it said, but it stops there. A majority can't make individuals pay for political parties.

The union's lawyers, civil libertarians Thomas Berger and Frank Scott, said the law was designed to keep unions out of politics, not to protect individual rights. It prevented freedom of association, took away the breath of life and controversy from public life. Union donations expressed a majority will, and didn't stop individuals from voting or acting as they saw fit, the union argued.

Ian McGilp, lawyer for OPSEU in the Lavigne case, says these 1962 arguments remain the key issues today. Lavigne and the NCC rely on the old court decisions, which saw unions as "forced," not democratic, associations, he says. If majorities can decide on bargaining, they can decide on political projects that promote bargaining goals, he says. "The Charter is supposed to protect democracy, and individual rights have to be consistent with democracy," he says.

Opponents of union political contributions have yet to criticize corporate donations made without customer or worker permission.

THE LEADER-PO[st]

Variable cloudiness

VOL. LIII—No. 154   FORTY-FOUR PAGES   REGINA, SASKATCHEWAN, SATURDAY, JUNE 30, 1962

MEDICARE TALK[S]

★ ★ ★   ★ ★ ★

Gov't pla[ns] [MD]

REGINA, SASKATCHEWAN, TUESDAY, JULY 3, 1962

VOL. LIII—No. 155   THIRTY PAGES

No solution in sight between doctors and government

MEDICARE EMERGENC[Y] MOVES INTO THIRD D[AY]

LAST EDITION   SINGLE COPY ... 7[¢]

VOL. LIII—No. 156   32 PAGES   REGINA, SASKATCHEWAN, WEDNESDAY, JULY 4, 1962

More doctors due from U.K.; 230 MDs work at centres

EMERGENCY CASES ARE DOWN TODAY

It is understood that so[me] staffed with medical personn[el] brought to Saskatchewan by [...] practising at Sha[...]

emergency-only medical [...] [mon]thly Wednes- [...]

Hous[e]
in T[...]

TORONTO ([...]) describing the[...] wives Tuesda[...] the Canadian [...] tion headqu[...] against the [...] tors, strike [...] ince's medica[...]
The group [...] reading:
"Why let [...]
"The CMA [...] and patient[...]
Several w[...] from Sask[...] have relati[...]

REGINA, SASKATCHEWAN, TUESDAY, JULY 10, 1962

Clergy trying to help find solution

MEDICARE CRISIS MOVES INTO COURT

Pews for cues

EDMONTON (CP) [...] [...] news for the use [...] a building in Edmonton [...] Ca]nad

City centre Monday night [...] [cele]brated the memory of St. Mary [...] [Cat]hol- [...] Serv]ices were offered a [...]

Focal point of Saskatchewan's medical care crisis Tuesday moved to the Regina courthouse where pro- ceedings that could suspend operation of the Medical Care Insurance Act could validity can be tested in court were adjourned until this afternoon.

Three Saskatchewan residents who seek to have the act declared invalid, asked for a week's adjourn- ment of their application for a court order suspending medicare commission activity until September. Counsel for the commission opposed their request

for an adjournment and asked that the application to suspend commission activity be dismissed. Argument on the issue was to continue before Mr. Justice D. C. Disbery at 2 p.m. Tuesday.

Premier Woodrow Lloyd and Dr. H. D. Dalgleish, president of the Saskatchewan College of Physicians and Surgeons, have indicated no break in the bitter impasse which has plunged the province into emer- gency-only type medical care.

Tuesday is the 10th day since most of Saskatche- wan's 750 practising doctors withdrew normal service

MEDICAL HISTORY

In May 1963 on the outskirts of Regina, soldiers stopped Bill Gilbey and a carload of unionists from the Co-op Refinery. An army guard scanned Gilbey's car for dangerous weapons. Then he passed out a leaflet denouncing medicare.

Troops don't ordinarily direct traffic in Saskatchewan. They don't usually check for guns. They don't often pass out political leaflets. But traffic was heavy that day, and the whole province was on a road to what seemed like civil war. It was a showdown over just what the doctors could order. The NDP government was pledged to introduce medicare on July 1. Doctors were dead set to block it.

The forces of organized medicine had numbers and power behind them. A on protest rally on May 23, for instance, boasted a cavalcade of 900 cars and carried 46,000 petitions from across the province. The waiting room of this medical resistance movement was filled with supporters from the social, business, and political elite across the country. The confrontation came to a head in July. Doctors struck against medicare for 23 days. They turned Saskatchewan into the battleground of public-health insurance.

Memories of the battle are still bitter, and military terms are still common. Even

263

those who aren't aware of medicare's stormy origins praise Saskatchewan as the "flagship" of public medicine. But flagships don't float in Saskatchewan. It's an inland province where the grass roots, not the admirals, set the political agenda. The medicare fight was won with ground troops. Saskatchewan power and refinery workers, both organized in OCAW, were on the front lines. They mobilized to safeguard the government's medicare program and to push it in new directions of community medicine.

If there was a place where medicare could be started up without fierce controversy, it was Saskatchewan.

Many doctors had been on the public payroll since 1915, when small towns guaranteed doctors an income if they ventured to the lonesome West. Just as doctors got used to being paid in cash, the Depression of the 1930s hit. Farmers tried to hold their heads high by paying off medical debts with chickens, eggs, and grain. Cash-starved doctors ate their pride and in 1933 came out for a government insurance plan that could pay their bills.

Painful memories of loved ones denied proper medical care, of the crushing debt of hospital expenses, topped the list of almost every Saskatchewan resident's medical history. The CCF, elected in 1944, set out to remedy the economic ills of a health-care system based on ability to pay. Canada's poorest province, groaning under the burden of bankruptcy-level debt left over from the Depression, found the money to pioneer public-hospital insurance, free cancer clinics, and expanded mental-health facilities. As soon as these programs became models for the rest of the country, the CCF set the pace for the next round of public-health expansion.

In 1960 the CCF-NDP staked its re-election in Saskatchewan on a promise to bring in government-run medicare. Doctors went flat out to defeat the NDP. But they couldn't misread the election chart. The day after the NDP's decisive victory, the general secretary of the Canadian Medical Association offered to co-operate. "This is democracy. The CMA accepts the decision in this light," he said. "Our efforts will be bent on avoiding the defects we see in government plans elsewhere."

That mood didn't last. Over the next three years, doctors turned down every government effort at compromise. Dr. E.A. Tollefson, a noted neutral in the dispute and later author of *Bitter Medicine: The Saskatchewan Medicare Feud*, couldn't figure out why. Saskatchewan's plan was closer to the CMA's policy than Alberta's, yet doctors embraced the Alberta plan, he said.

Tollefson guessed that the irrationality was caused by the hysteria campaign waged against medicare. By the time the government was ready to start its plan, doctors had worked themselves up to a fever pitch. Egged on by the entire business and social establishment, they became shock-troops in a campaign against the very principles of social security.

The medical profession had become so closed and tight that it lost its ability to deliver a second opinion. Although Saskatchewan was dangerously underdoctored — across Canada one doctor served 879 people, while in Saskatchewan one doctor had to look after 1,019 people — doctors were cut off from the hard lives of their patients. By 1960, according to Robin Badgley and Sam Wolfe in their book, *Doctors' Strike: Medical Care and Conflict in Saskatchewan*, only half the province's 750 doctors had experienced the lean years of the 1930s and 1940s. Over half were cloistered in Saskatoon and Regina, forming a close-knit network kept in isolation from three-quarters of the population. A third of Saskatchewan's doctors were "political refugees" who had fled post-war Britain to escape medicare there. Throughout the 1950s they nursed their bitterness with scare stories about socialized medicine in the province's professional journal.

Home-grown backers of the medical establishment, organized in the Keep Our Doctors Committee (KODC), rallied all the opposition elements who wanted to perform radical surgery on NDP accomplishments in the province. THE KODC kept up appearances as a movement of housewives, teachers, and ordinary citizens who wanted the government to back away from a confrontation that would force doctors to leave the province. But according to Wolfe and Badgley, KODC backrooms were dominated by business groups that "kindled prejudices about basic values such as freedom, nationality and class, and converted these into subtle political issues."

Father Athol Murray, a leading light of the KODC, wasn't that subtle. He told one rally: "Tell those bloody Commies to go to hell when it comes to Canada. I loathe the welfare state and I love the free-swinging freedom. I am seventy and I'll never ask you for the Old Age pension — to hell with it — I want to be free." A U.S. television reporter, fresh from a stint covering violence against civil rights marchers in the deep South, said of one KODC march: "This is just like covering the anti-integration movement in the southern United States. They're the same kind of people."

The Saskatchewan media, without exception, blacked out any support for medicare, played up the KODC, and backed the doctors' strike. An editorial in the Moose Jaw *Times-Herald* on July 1 was headlined "The day that freedom died in Saskatchewan."

But media from outside the province put the opposition in a national focus. If the doctors' strike succeeds, "the cause of adequate health insurance in North America may be set back by as much as a decade," the *Toronto Star* argued on July 6. "That is why reactionaries, in and out of the medical profession, are rallying to the support of the striking doctors -- including many who normally oppose all strikes."

265

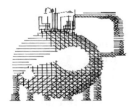

Premier Woodrow Lloyd liked to speak quietly and carry a small stick. He followed what he called a "strategy of silence," the political equivalent of bedside manners designed to maintain public calm and make it easier to re-establish friendly relations with doctors once the strike was over.

Lloyd had no intention of backing down from a confrontation. On June 24 he authorized a contact in England to organize an "airlift" of doctors willing to fill in during the doctors' strike. But Lloyd also had no intention of heating up the confrontation. His daughter says he was deeply hurt by an editorial in Biggar, his home town, that railed against "some strange gook from a foreign country" invited in for the NDP's airlift. It unnerved him that such ugliness could be unleashed by conflict. He asked supporters to stay shy of staging pro-government rallies that might provoke the opposition.

Members of the Oil, Chemical and Atomic Workers didn't quite follow this prescription for silent partners. In collaboration with the Saskatchewan Citizens for Medical Care Committee, the Community Health Services Association, and other members of the union and co-op movement, they made sure the pro-government silence wasn't too deafening. By organizing grassroots support and setting up community clinics for "strikebreaking" doctors, they helped stiffen the government's resolve and point medical insurance towards new concepts of medical practice.

As the premier union in the province, OCAW was largely responsible for the Saskatchewan Federation of Labour's 1962 brief to the famous Royal Commission on Health Services, chaired by Saskatchewan-bred Emmett Hall. At a time when most people saw medicare as an insurance scheme against sickness, the Federation insisted on co-ordinated health-care services that ensured a focus on prevention. "To attempt to devise a program of health care with the objectives of promoting health, preventing disease, rehabilitating the ill as well as caring for the sick is altogether a nobler purpose," the brief said. It was signed by, among others, Mike Germann, president of OCAW's Canadian Council and vice-president of the Saskatchewan Federation of Labour.

Saskatchewan Power workers metered more than electricity during the doctors' strike. The local union headquarters metered mail for one of three doctors in the province who openly suported medicare. One Saturday night the medical loner did a mailing of his nine hundred colleagues, courtesy of the OCAW. When a medical alert team tracked down the code on the meter, the Canadian Medical Association went for the jugular. "More evidence of union infiltration and control of the present government," it claimed.

As anti-medicare opinion spilled over into anti-unionism, the media found a poster boy in Hans Taal, financial secretary of the Saskatoon Power local and vice-president of the KODC. He was allegedly told by his union rep, Buck Philp, that he couldn't ride two horses at once. Taal quit the union and his job to work for the KODC. The press had a field day. "This creates the impression that the union bosses consider themselves to be a law unto themselves, above and beyond the guarantees of" human rights, an editorial in the Saskatoon *Star Phoenix* stated.

OCAW leaders were key players in Regina, Saskatoon, and Estevan community clinics, set up to house doctors airlifted from Britain who treated people during the strike. Sask Power local president Reg Basken chaired the Estevan co-op. Basken says his memory is still "very vivid because my son was born in May and it was very hard to get a doctor to look after a complication with his blood. That just strengthened our views that honest medical care was required, and we participated fully in all of the medical battles that ensued." Sask Power vehicles drove patients to Winnipeg and Calgary when emergency hospital treatment was necessary.

On July 11, the KODC overshot its mark when it staged a mass anti-government rally in Regina. Most drugstores and many companies gave their employees the day off to attend the protest. Some forty to fifty thousand demonstrators were expected. Less than five thousand showed up. The doctors lost heart and started to negotiate. Lord Taylor was brought over to mediate on July 16, and got a face-saving settlement for the doctors on July 23. The government medicare system remained intact, but individual doctors got the right to opt out.

Taylor demanded his payment — a week at a northern fishing lodge. Then he ordered the province to undergo a period of "absolute rest" from controversy. Doctors went back to work. The death rate in Saskatchewan had actually decreased by 38 per cent during the strike — one baby died in his mother's arms while awaiting transport to an out-of-town hospital — but it seems most people had their lives extended by the closure of hospital operations. Once doctors caught up, they did well under a government insurance plan. By July 1964 there were 124 more doctors in Saskatchewan than before the system that was supposed to bring medical doom and gloom. By 1965, according to Geoffrey York's *The High Price of Health*, 72 per cent of doctors said they were pleased with the new system.

Maclean's named Premier Woodrow Lloyd the outstanding Canadian of 1963. He "has made his revolution and he's made it stick," the magazine said. The first medicare plan on the continent became established, laying the basis for a Canada-wide plan later in the decade, introduced without a major confrontation with doctors. But "revolution" and "absolute rest" didn't co-exist easily in Saskatchewan.

Medicare was no cure-all for the problems of traditional medicine. Doctors accepted payment from the government but beat back any

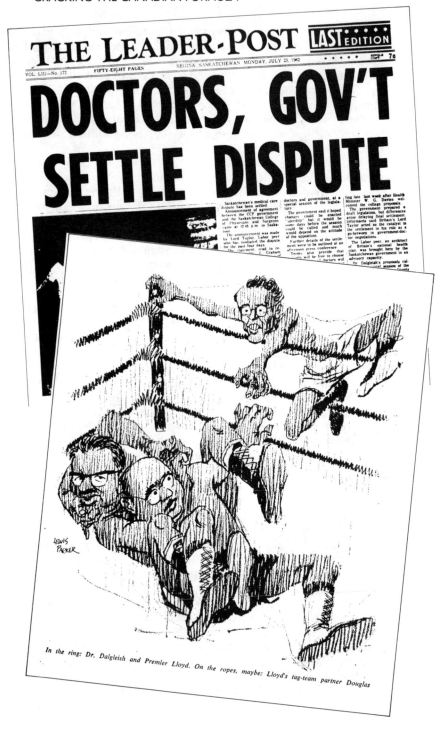

In the ring: Dr. Dalgleish and Premier Lloyd. On the ropes, maybe: Lloyd's tag-team partner Douglas

strategy to use medicare as an alternative to the traditional solo practice of fee-for-service doctors who treat illness rather than promote health. Even though governments picked up the doctors' tab, medical associations retained the powers of "private governments," according to Malcolm Taylor's *Health Insurance and Canadian Public Policy*.

The community clinic movement, born in the struggle against organized medicine, lost out when doctors adapted to medicare.

Doctors who staffed makeshift community clinics during the strike were denied hospital privileges by local medical elites. The medical association refused to allow clinics to advertise for members, charging that such advertising was unprofessional.

In 1963 the Canadian Council of OCAW donated $500 to help clinic doctors with their struggle. The Sask Power local donated $1,000, stating that "to win medicare, we have to win hospital privileges" for clinic doctors. Mike Germann led an SFL delegation that condemned the medical association for denying patients of clinic doctors entry to hospitals. "It's the same as taking away patients' rights to choose their own doctor," he said — the very thing doctors accused medicare of.

But the need for absolute rest after such an intense struggle prevailed. The NDP was defeated in the next provincial election. Even the popular Tommy Douglas, the first CCF premier of Saskatchewan, was defeated when he ran for a federal seat in the province.

The relapse after the medicare controversy made it difficult to introduce more basic reforms, a 1974 Science Council report on science and health services noted. Only seven of the group practices and community clinics survived. "Those which did provided a major contribution to the Canadian experience in this mode of operation, which is again attracting much attention all over the country."

That report was also too far ahead of its time. Community-based, prevention-centred medicine didn't become a catchword until the late 1980s, almost 25 years after it was raised to the fore by Saskatchewan workers.

THE SECOND COMING OF SOCIAL CREDIT

In the Alberta of the 1930s, Premier William Aberhart, a famous radio evangelist and the great communicator of his day, focused Depression desperation onto his scheme of social credit, a promise to put Monopoly money in the pockets of the poor.

His feisty supporters wanted that and more. They thought nothing of working with Communists, socialists, and unionists to get it, Alvin Finkel notes in The Social Credit Phenomenon. Though Aberhart never did bring in social credit, he made Alberta a national leader in labour, social, and health legislation: the first minimum wage laws for men, compulsory overtime pay, the first labour laws recognizing unions, free specialized health clinics.

When Aberhart died in 1943, power passed to his protégé Ernest Manning, host of Canada's "National Back to the Bible Hour." Manning turned the party from social credit to social conservatism. The world was still divided into good and evil, the will of government was still the word of God, opposition was still conspiracy, but the new enemy was socialism, defined as a front for the world banking conspiracy.

"From 1944, the struggle against socialism practically replaced the struggle against the financiers, while being represented as the same thing," C.B. Macpherson wrote in his classic book Democracy In Alberta. Social Credit "had come to rest in the position of an orthodox conservative government."

The oil boom completed the transformation from wheat-belt rebels to high-octane reactionaries. But Social Credit still provided a home for the lonely and the marginal who ached to belong, William Mann wrote in Sect, Cult, and Church in Alberta. As statistics in Martin Robin's Canadian Provincial Politics show, Alberta spent more than other provinces on social services: Revenue from oil royalties made financing easy. Alberta had no sales taxes.

Thus Social Credit had the best of all possible worlds. It milked the insecurity and paranoia of a devoted following that felt left behind, victims of external forces. It left emotional wounds raw, catered to their pain and anger with righteous indignation and a promise of individual salvation. It

▼ PART II

▼ SECTION 2 / BRINGING IT ALL BACK HOME

▼ CHAPTER 3

soothed immediate economic hurts, pre-empted opposition based on the need for specific government programs. It dealt quietly and generously with multinationals.

No other government in Canada enjoyed the combination of windfall prosperity, active support from the business establishment, worshipful reverence from a mass following, and apathy from the rest. It was scary. It was held together by the Red Scare, by the dread of eternal hell-fire if government social programs damaged individual salvation, by the fear of economic collapse should unions ever burst the bubble. Like all governments of totalitarian bent, Alberta's pictured unions, a force for independent thought and action, as a source of unrest and subversion.

FEAR AND LOATHING ON THE CAMPAIGN TRAIL

Alberta is the badlands of Canadian politics.

The province has wheat, cattle, coal, and oil, but no cushions. There's no tomorrow if the prices of farm and energy products fall, no middle ground to block a fell swoop of yesterday's politicians. Voters made a clear swath of federal Tories and provincial Liberals after the First World War, replacing them with Progressives. Social Credit trounced them during the Depression by promising an all-out fight against the Eastern bankers and their political hangers-on. The province switched from one-party Socred government to one-party Conservative government in 1971.

In the late 1980s, New Democrats got their turn in "Redmonton." The blue-collar capital of the province gave them victories in federal, provincial, and municipal elections. In 1989 Jan Reimer won every poll in the city to become the country's most straight-talking, radical mayor. The politicians couldn't figure how she stayed calm in face of shrill attacks, how she turned the tables on hecklers, glad to be taken for what she was. Her campaign chairman, Neil Reimer, stopped worrying about the political bruises his daughter had suffered in childhood when he was NDP leader and she was attacked by students, teachers, and anonymous phone calls.

In 1967 Jan Reimer had peered out a motel window onto the main street of Edson and watched a group whooping it up over the election shut-out of the NDP. A mock funeral train carried a casket bearing Neil Reimer's name to its political grave.

The service had been conducted earlier in the week. In a show of last-minute respectability, Reimer went to church. The minister asked everyone to join in prayer against the ungodly socialist. "That's the first time that a guy's got me up and had me actually praying against myself," says Reimer. "The Lord must have heard my prayer, because I was defeated."

Jan Reimer is one of a few new-wave New Democrats with training that goes back to the 1960s, the dinosaur age of Alberta politics, when the fear of God was put into voters, when New Democrats were infidels. "There was a fear to be anything but Socred and a fear to discuss politics in general," Jan Reimer says. "It's that simple."

As Alberta NDP leader for most of the 1960s, Reimer tried to get a political toehold by pleading for an "open society." A modest theme today, but subversive then. Alberta "was a very closed society," he says. "You were either Socred or nothing. And I think there is some real challenge in opening a book. The pages of the book were all stuck together and we had to unglue them. And that's not an exaggeration."

Jim Coutts, later Pierre Trudeau's political secretary, was an Alberta Liberal in those days. He says Social Credit mixed "prairie progressivism and right-wing boosterism in the same party," casting a wide net and playing both ends against the middle. "They positioned themselves on the far right, but they also kept up an image as Big Daddy looking after the people," Coutts says. They made Liberals and Conservatives look too conservative and New Democrats too radical. Reimer didn't have a chance, according to Coutts. "It was bad enough that someone should be a socialist, but to be a unionist too, when unionists came from cities in the East, was a double curse."

Reimer never saw his union job as a cross to bear. He says it took an "untouchable" to take on the Socreds the way they had to be taken on. "It had to be a unionist to get the party off the ground in Alberta, or somebody very rich," he says. "Lawyers were afraid to touch us in case it drove business away. Doctors wouldn't touch us after the medicare dispute. A small businessman would have been driven out of business." Unions pay their dues to society by being "a sanctuary, where you can stand up and speak up about the ills of society as long as you have union support, and not worry about putting bread on your table or being run out of the province."

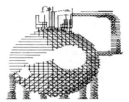

Reimer took a quick dislike to Alberta's Social Credit. When he went from Saskatchewan to Alberta as a union rep in 1951, he was used to laws that gave unions a break. He was used to debates where devil's advocates were guests of honour. Alberta's labour laws favoured company unions and did him out of a few locals he thought he deserved. The conformity, self-righteousness, and narrow band of tolerance in political discussions smacked of thought control.

In 1964 Reimer clipped an editorial from the *Calgary Herald*. There is no "free play of ideas," it complained. Social Credit "is nothing more than a form of corporate authoritarianism administered by infallible leadership." Later still, a 1989 history — *The Social Credit Phenomenon in Alberta* by Alvin Finkel — summarized the party's legacy as "obscurantism, xenophobia, and the extolling of greed in the name of religion and liberty," all mixed with a "pretence that Albertans speak with one voice."

But in the 1950s it took an outsider to feel the snap of authoritarianism in the air, the cloud of suspicion imposed by the political guardians of the moral majority. Even among union leaders Reimer knew he was up against more than a government, more than a political party with a particular point of view. He was up against Big Brother — or more accurately, Big Father. Union leaders in the province were squeamish, afraid to speak out. They bent their knees and doffed their caps to government leaders, as much as any group. They prefaced any complaints about the province's labour legislation, the most anti-union in the country, with vows of loyalty to the battle against subversion. Alberta's labour acts of 1948 and 1960 were the most harsh in the country but unions were afraid to raise more than modest objections, lest they become whipping boys.

When Premier Earnest Manning was invited to give the keynote speech at the founding convention of the Alberta Federation of Labour in 1956, his theme was the horn of plenty that Alberta had become, that labour could share in. The next day Reimer spoke about his boyhood days and about watching newborn pigs in the barn. One sow had 12 tits and 13 piglets, he said. "There was a horn of plenty in that sow, but that 13th piglet is like labour under Manning. We're continually looking for that tit." The chair cut him off.

273

Throughout the 1950s the Socreds faced no serious opposition.

Provincial Liberals found political meaning in federal patronage, and the federal Liberals had a non-aggression pact with a provincial government it couldn't afford to alienate. The CCF fell into place, last place, as the socialist counterpart to Socred fundamentalism. The two parties were meant for each other, both products of a prophetic, other-worldly political culture. In the 1944 election, when the CCF had a chance of toppling the government, it ran on a straight-up, one-plank socialist platform.

There was no use campaigning for specific reforms, Alberta's CCF leader Elmer Roper explained, because the Socreds had convinced people they couldn't be achieved. In the 1950s the CCF concentrated on foreign-policy issues that challenged the cold war. The meat and potatoes issues of provincial politics didn't interest the party. It didn't even organize provincial constituencies.

The CCF saw little need to organize. According to Gad Horowitz's *Canadian Labour in Politics*, CCF leaders whiled away the eve of the 1959 Alberta election playing bridge. They told a visitor that the masses would come to socialism when they were ready. They weren't ready that night. The CCF got 4 per cent of the vote, down by over 20 per cent from what it got in 1944. Roper, the party's popular leader, was defeated. It was "the lowest point in our history," says Ivor Dent, a CCF warhorse who later became mayor of Edmonton. Roper "got defeated because these people did not see that in fact poll work means sweat, and you bloody well do it or you don't win.

The Alberta NDP's 1962 founding convention was a shotgun marriage between old-time CCFers and New Party clubs sponsored by unions and federal party organizers. The CCF was a reluctant bride. The last CCF convention in 1961 refused to "irrevocably commit itself at this time" to the NDP. Members thought the new party was too open to "subversion" by Liberals, Conservatives, and narrow trade unionists. The provincial executive voted to keep its own identity as the Woodsworth-Irvine Fellowship.

Reimer was elected president of the Alberta NDP, an unpaid position that is usually an organizational job. But in Alberta, with no NDP seats in the legislature, the president was in effect the leader of the party — a full-time position in any province. Meanwhile, he hung on to his job in the union.

Reimer wasn't a gracious groom. He tried to bar entry to the Woodsworth-Irvine Fellowship members until they made a solid commitment. "They didn't turn a goddamned thing over to the party, not even a worn-out typewriter," Reimer says. They did turn over five hundred membership cards, half of them carrying the names of dead people.

The odd couple bickered over everything. CCFers saw Reimer as a dictatorial, right-wing union hack. Reimer saw them as "nothing but a debating club." This trade of insults between die-hard CCFers and unionists has echoed throughout NDP history. Big-labour power brokers and compromisers, against grassroots principles; or unrealistic purists, heads in the past, against serious politics, reaching out to new voters.

Academics trying to find a pattern in this rhetoric have concocted a theory that the difference was between those who wanted to build an idealistic movement and those who wanted to build a modern political party of the centre. In Alberta, the two factions were further apart than that. They couldn't agree on the difference between idealism and naivety, between a movement and a cult. There was no tug of war, no creative tension. They just ignored each other.

Reimer was an old CCFer himself. He was not more "left" or more "right" than the old-guard CCF. But he came to the NDP as an unrepentant union organizer. He was keen on new members. He was an agitator, a hell-raiser. He looked for issues, not platforms. He knew that majorities had to be won over, not added up, that the key to an organizing drive was a solid core of activists, not a polished speech or an abstract vision. That's why his work for the NDP didn't interfere with his continuing union duties as Edmonton staff rep with the OCAW. Shortly after he became NDP leader, a worker dropped into his union office. "I want to join that Reimer-union," the man said. "We need someone to raise hell in our plant." Reimer handed him the organizing cards. As the man left, he asked, "By the way, what's the name of that union?"

Reimer says he organized six plants that way. "If you're out there taking on an injustice, workers will identify with you," he says. In fact, Reimer says he allowed himself to be roped into the job as NDP leader because "I thought we would just be organizing, and not taking very many public positions. That likely shows my naivety about politics at the time, because in the absence of a leader, the president is the leader."

As an organizer, Reimer wanted a clean two-way fight between the NDP and Socreds. That meant knocking out the Liberals, positioning the NDP as the only alternative. The Liberals had the same designs on the NDP, and began creeping up on the NDP's left. They came out for public ownership of electric power plants just as the NDP was about to meet. Reimer opposed outbidding the Liberals', said it was enough to control the distribution system, that the money could be better spent buying public control of the oil sector. He barely kept the convention from backing wholesale nationalization without compensation.

The program was designed to look moderate, he says. "This was the single most important issue that made the Liberals fall between the cracks. I didn't bite their bait. So it gave us the first bell to ring in a way that people would listen to its chimes. All of a sudden I became credible. Had we gone for a radical position on a side issue, we would never have got off the ground because those first impressions last."

The subtlety was lost on most NDP members. Reimer did not want to build a moderate party of the centre. Nor did he want a program that put the party on the left fringes. He wanted to forge a fighting party, one that had a fighting chance of leading average people in struggle. That's the unique perspective of a union organizer, alien to party professionals and ideologues alike. Reimer tried to straddle the two poles, and in the end satisfied neither.

Reimer psyched himself up to shock the membership into optimism. He told the founding convention that the NDP would have ten thousand members within the year. Unless workers affiliated through their unions were counted, he was off by 7,300. The NDP didn't attract anyone interested in a political career. It wasn't a place to get ahead, not even to meet people. The party had to rely on people of character, and it got built by characters. Ted Chudyk, a former Ukrainian youth and farm organizer, was the party's legendary fundraiser. He had the makings of a professional gambler or high-stake entrepreneur and had hundreds of schemes to bring in new members and donors. Grant Notley, the inexhaustible party secretary, drove around the province organizing constituency clubs. His trademarks were threadbare coats, worn shoes, and his own version of "the people's flag is deepest pink." The party was built out of sheer will and bluff. In the 1963 provincial election Reimer and Ivor Dent put mortgages on their homes to finance the campaign. In 1967 Basken did the same. Their wives didn't find out until many years later. Reimer set a modest goal for the 1963 provincial campaign, which followed shortly after a federal election. "Our function was to come out of it with a party after the election," he says. Premier Manning was more ambitious. His campaign slogan was "63 in '63." He wanted to win every seat.

Reimer and Notley beat the bushes to find candidates for every riding but first had to organize riding associations that would nominate them. They twisted arms to get candidates. In Grande Prairie, for instance, three people were nominated, two declined, and Reimer jumped up to delare it unanimous for the third, before that person had the chance to decline. A number of candidates, mainly young people, were parachuted in. The average age of NDP candidates was 38, of Socred candidates 56.

Reimer was still awestruck by the closed nature of Alberta society. One farmer he canvassed took him behind the barn and slipped him 20 dollars. "I don't want my neighbours to know and anyway I can't vote for you," the farmer told Reimer, "because Manning is a man of the cloth and the Lord may cause a terrible visitation upon us."

The difference between Saskatchewan and Alberta led Reimer to run with the theme "Let's create an open society." The NDP "will make democracy in our province a meaningful and developing process," Reimer promised. He attacked the Socreds as smug and power-hungry. Manning had too much power. He was premier, attorney-general, and minister of mines and resources. He treated the legislature as a rubber stamp. Reimer charged that Manning had passed two thousand laws as cabinet orders-in-council, thereby avoiding the legislature.

The NDP, Reimer pledged, would "restore the balance" and "foster an increasing measure of economic democracy and responsibility in Alberta." Restoring balance was "the top issue." It meant unions and co-ops strong enough to take on the giant corporations. It meant medicare, car insurance, and crop insurance funded by government. It meant provincial use of oil royalties to fund education and take the tax load off muncipalities.

The NDP got 10 per cent of the popular vote, twice its 1959 standing, but elected no candidates. Reimer himself lost out in North Edmonton by 1,500 votes. The *Edmonton Journal* declared the election a "smashing victory," stating "There is little doubt the Manning image, a protective shroud which resists attack, is the key to success."

At the post-election council meeting of riding leaders in June 1963 Reimer put the best face possible on the defeat. He said party membership had climbed from 2,695 in January to 3,246 in April, making Alberta the fourth strongest NDP province in the country. Only Saskatchewan, Ontario, and British Columbia were ahead. He said success in the future depended on hard work and not hardline policies.

"From this election, one can clearly see that the biggest need of the party is not the need of adjustment in our policies," he said, "but concentration and the realization that politics is work, and we need workers and organization."

Reimer's speech wasn't just a pep talk for more of the same. He told the council that the party had to make a sharp turn, had to take off the gloves. "It was clear to me that everybody was afraid to attack Manning," he says. "How long can this continue? If you take the attitude that he's a man of God and that we can't attack him for his economic policies, he'll be in there forever."

That remained Reimer's refrain for the rest of his term as leader. It was the only way he saw of breaking the Socred hold on public opinion. "We will not defeat Social Credit until we get the people in a questioning frame of mind," he told the provincial council in 1965. "Once the public grows aware of the shortcomings of Social Credit, and begins to search for an alternative, then I believe we must be ready with a positive program. A good program is not enough, however. We must train our people to work politically."

Reimer started lashing out at Socred corruption in 1964. He attacked cabinet minister Alfred Hooke for accepting a free stay at the Vancouver apartment of Dr. Charles Allard, the prominent surgeon and landlord. Allard later got a cheap rate on a $1.5 million loan from the provincial treasury. This knocked Hooke off the high horse he rode in November 1964 when he lashed out at "trash" and "moral decay" in the schools, a reference to sex education. Reimer went after cabinet minister Dr. Lou Heard, accusing him of conflict of interest in refusing to recognize chiropractors.

Reimer drew blood when he nailed cabinet minister Edgar Hinman for mixing private and cabinet business. Manning fired Hinman in July. The *Edmonton Journal* called the Hinman scandal the "biggest bombshell of the past decade" — but no bomb went off. Manning refused an investigation. "I don't think the people of this province have any right to know any more than they have been told," Manning said.

Reimer knew he was appealing to a puritanical strain common among hardworking people who hate hanky-panky and political privilege. He tried to lend a socialist perspective to his inquisition. "This is not surprising that we have this kind of climate in Alberta," he told the 1964 NDP provincial council, "because after 25 years of a philosophy of exploitation, a philosophy of `me first and to hell with the unfortunate and the hindmost,' we are bound to have the climate we are experiencing today."

The NDP was on a roll. The press began referring to Reimer as the unofficial leader of the opposition. A 1965 byelection in the mining and rail town of Edson gave him his chance to make it official. Edson was the heart of a riding that had elected Social Creditor Joseph Unwin to the legislature in the 1930s. In 1937 Unwin was sentenced to three months of hard labour for branding nine prominent Edmontonians as "bankers' toadies" and urging the public to "exterminate them." Reimer crafted his appeal to make the most of that tradition. Social Credit was once "a party that professed to protect the people from the monopolies, but has now changed to a party that protects the monopolies from the people," he said.

Election leaflets urged people to "vote for the man they can't ignore. One good man in opposition is worth ten yes-men on the government benches." *Maclean's* magazine assigned Arthur Hailey, the Canadian novelist and an acknowledged conservative, to check out Reimer's charges. Hailey pronounced Manning to be genuinely religious and free of corruption but agreed the premier had been lax. Hailey described Reimer as a constant thorn in the side of the government and conceded that the NDP leader would be "an asset to the legislature."

Reimer got 1,902 votes, almost triple the NDP count in the 1963 election. He lost by 93 votes. For the first time in a decade, victory seemed possible.

Reimer kept up the attack. It was working. He enjoyed it. In 1964 he accidentally climbed on board a plane carrying the entire Socred cabinet. He sent a note to the pilot. "If this plane starts to go down, don't save it on my account." Manning couldn't understand why a stewardess broke out in laughter when she passed his seat. In 1965 Reimer broke the story of a government land giveaway to private developers who had bought a parcelled lot for 25 cents an acre and sold it for $2,000. "On the one hand there is a giveaway to foreign interests, on the other extortion from Canadian workers," Reimer told the NDP council in March. Even Manning got into the spirit in 1966. By this time he was thinking of moving into the national arena, of launching a new and truly conservative party that would attack the NDP frontally. In October 1966 Reimer came up with more dirt on Hinman. Jake Superstein, one of Hinman's former partners in an airport strip development, had got burned on an investment and started to name names. Testifying that political fees had been charged for government loans, Superstein implicated the former minister of resources, Nathan Tanner, and said corruption was involved in another government giveaway to a major forest products company.

That's when the NDP hit political paydirt. That same month New Democrat Garth Turcott won a by-election in the Crow's Nest Pass area. Ted Chudyk, the campaign mastermind, pioneered the intensive door-to-door canvass, ever since a standard feature of NDP election organizing. It took four rounds of door-knocking, Chudyk says. The first time at a door the reaction tended to be one of disbelief. On the second time canvassers were told to buzz off. The third time people said, "You guys are really serious." The fourth time many of them said, "We're with you."

The upset signalled a political revolt, dissatisfaction with all old-line politicians. A random survey during Klondike Days, Edmonton's summer celebration, recorded a stampede away from the Socreds. The government wasn't doing enough about foreign ownership, said 58 per cent of those surveyed. The Socreds were a one-man show, said 28 per cent. Only 32 per cent said they would vote Socred.

Tory strategist Dalton Camp toured Alberta in October, and told the *Edmonton Journal* that the NDP could win if the old parties didn't shape up. Manning had the same worries. "The NDP has gained a foothold in the Legislature of Alberta because the free enterprise vote was divided between the Conservative candidate and that of the government party," he wrote a Tory MP. "Isn't this a situation that we should try and avoid if possible?"

Manning had already scheduled a short housekeeping session to follow the by-election. Garth Turcott, a shy and retiring country lawyer with no political experience, was immediately thrown into the spotlight. Grant Notley and Reimer tutored Turcott on the airport affair and told him to needle the premier. They sat in the gallery as Turcott read from the press reports and asked if the premier would launch an inquiry. No, the premier said. Turcott repeated his question. The premier repeated his answer.

Reimer and Notley left the legislature. They were ecstatic. They had visions of a media field day, of a prolonged period between sessions when they could hammer the government for refusing an inquiry. Then they heard a radio bulletin. Turcott had kept on pressing the premier, asked why it cost more to pave roads in Alberta than Saskatchewan. The minister of highways raked Turcott over the coals and the premier blew his stack, accusing Turcott of lowering the dignity of the legislature. The attacks had apparently gone too far, cut too deep. The press was not amused. Turcott was Reimer's Charlie McCarthy: The voice was Turcott's "but the hand holding the dagger of slanderous innuendo" was Reimer's, the *Edmonton Journal* said. After one legislative punch-out, Turcott fainted from the stress.

Reimer whipped up his followers to press the offensive. In February 1967 he told the *Edmonton Journal* to expect "the greatest political upset in Canadian political history." In March he told the provincial NDP convention that membership had increased by a thousand in the last three months, that the party had an election war-chest of $55,000, that 20 or 30 seats could be won. The Socreds scrambled to push Reimer's charges to the sidelines. Manning appointed an inquiry into corruption on April 7, ten days before announcing the election. Charles Allard sued Reimer and the CBC for slander, forcing both to remain silent on his sconduct during the election.

The NDP came on strong. It nursed resentment against high-handed government and corruption. It also highlighted a positive 14-point program featuring a provincial takeover of educational expenses. That demand had enjoyed widespread backing since the late 1950s and was supported by many rapidly growing but cash-starved towns and cities. The NDP campaign also stressed union rights for civil servants, and public auto insurance. Phil Thompson, the first Native candidate in the province's history, promoted the issue of Native rights.

Historian Alvin Finkel believes the Socreds had a mole planted high up in the NDP. Manning somehow got the NDP's advance plans, and eight days after the election was called he was able to print a Socred pamphlet responding to NDP charges. The pamphlet hammered NDP proposals to impose higher royalties on the oil companies and to place the tar sands under public control.

The NDP campaign ran out of steam in the last week. Reimer's speech at a leaders' debate on May 17 fell flat. It had no punch, no zip. "Only if we move towards a contemporary social democracy in which policies are determined by the legitimate needs and aspirations of free men and women with minds of their own can Alberta fulfill its destinies," Reimer mumbled. He looked like he was reading the speech for the first time, the *Edmonton Journal* said, and in fact he was. He had tried to prepare his candidates for the hard-hitting campaign. "The water's going to be cold," he told a candidate's training session in Red Deer. "Don't try to turn around in the middle of the river."

But when the pressure heated up, Reimer says, "the water did get damn cold. Many of them didn't want it to get above their ankles." He was pressured to turn himself into a statesman. "I knew there wasn't time enough for that," he says. "I should have given it another shot. People were waiting for the final shot."

But he gave in. "I had the material, on high level corruption, interference in the courts, and media patronage. I should have gone in for the kill." On the last week of the campaign he felt the campaign going down. "It was the most helpless feeling of my life," he says.

Against his better judgement, Reimer also agreed to run again in Edson. He wanted to run in Edmonton, to capitalize on its large blue-collar vote and escape the enforced conformity of a smaller town. Ivor Dent came within a hundred votes of winning Edmonton, despite the fact that he was in Thailand for the entire campaign.

The Edson riding was huge, stretching to the Rocky Mountain pulp town of Hinton. Manning made the riding a priority, and came on the eve of the election to call for a united opposition behind the Liberal candidate. The Socred vote collapsed, the Conservatives didn't field a candidate, and Reimer lost to the Liberal.

The breaks went to Peter Lougheed's Conservatives. Bright, energetic, urban, fresh, and modern, Lougheed shone at the leaders' debate on May 17. He wasn't interested in slinging mud, he said, but in the shortage of nurses and fresh ideas. He dazzled the TV cameras, then a new fixture in Canadian politics. He was "an image looking for a cause," writes oil industry analyst Peter Foster. "The cathode ray tube had finally conquered the Bible in Alberta."

On election day in May 1967, the NDP netted 16 per cent of the popular vote across the province — almost double its previous run — but was wiped out. The Conservatives captured 26 per cent of the vote and got six seats. The Liberals got three seats with 11 per cent of the vote. Social Credit's popular standing fell from 54 to 44 per cent, but it got the lion's share of seats. "Albertans have handed the liberal and left-wing total welfare staters a resounding rejection," the *Edmonton Journal* cheered.

In his post-election analysis for the *Alberta Democrat* magazine Grant Notley blamed the Canadian personality for the NDP failure. "It seems to me that the greatest shortcoming of politicians, or Canadians as a whole, is the 'play it safe' or the 'easy way out' philosophy," he wrote. "Whether you talk about our uncritical acceptance of foreign capital, our indifference to the plight of the underprivileged both at home and abroad, the `play it safe' conventional wisdom contributes to a 'grey sickness' which permeates Canadian society." There were no funds to publish this post-mortem until January 1968.

The Alberta NDP was broke. It had spent every last cent in the campaign. It went in debt to hire lawyers for the inquiry set up by Manning to investigate corruption. Yet it needed a full-time leader to take it forward. Reimer had never taken a cent from the NDP and wasn't in a position to quit his union job. He also knew he couldn't shake the negative reputation he'd built for himself. Shortly after the election he announced his intention to resign at the November 1968 convention.

Privately, he had come to the view that provincial politics was irrelevant in the age of multinational corporations. "I thought that the time for genuine social democratic governments had passed us by, that political ideology was smothered by international events," he says. "It was my view that only one factor could have an impact and that was international labour organization." This was his private view of things — he did not share it with party activists at the time.

Reimer gave a scrappy farewell address. He said he was avoiding the temptation to give a pleasant speech. He had set certain objectives for the party, the top one being to re-establish political dialogue in the province, to show no government was above reproach. Someone had to do it, and no one else would. He also wanted to make the party relevant, to become a real opposition, though "not so far left so that we wouldn't have a practical meaning." He wanted to overcome the CCF's reputation as a clique that just talked about foreign policy. Politics isn't just a matter of discussing different programs, he said. It's fighting for a point of view.

He welcomed the youth revolt of the time and asked the party to identify with its spirit, to go beyond reform of taxation, medicare, and oil policy. The NDP "is needed for things much more fundamental than these matters of program," he said. It is "needed to improve the quality of life" and "for the dignity of society, for the dignity of humanity and the individual." He asked the party to concentrate on these basic problems of industrialism, to seek the answer in direct democracy, not just higher wages.

Manning resigned from provincial politics the same month, still hoping to establish a true party of social conservatism that would confront socialists head-on. The bloom had been taken off the social conservative rose a month earlier, when W.J.C. Kirby released his report on provincial corruption. Although Kirby couldn't prove the NDP's charges, he agreed the ministers had been lax and "imprudent." Everyone knew they were richer because of it. The fall from grace, the feet of clay — together with the slip in political unanimity — denied right-wing forces the distinctive moral authority, confidence, and identity needed to galvanize a populist base behind conservative economics. The movement to build a national party of the far right was stopped where it began, in Alberta.

It's hard to gauge the significance of that, save by glancing south of the border, where great communicators simplify complex issues, where personal piety is a political qualification, where profit is sacred and social legislation is sacrilege, where individualism and self-reliance mean conformity and private enterprise, where profound social questions are evaded with moral platitudes, where simple reforms like medicare are branded Communist, where rage and distress are focused on external conspiracies, where the very language of politics stifles debate, where ordinary people don't own or control their own anger. In Canada, even the Conservatives, lacking the fanatic resolve of a dedicated ideological faction, admit universal social programs like pensions and medicare are a "sacred trust." That didn't just happen. The door to the far-right had to be closed.

Manning left in time to avoid Social Credit's collapse. His successor, Harry Strom, was trampled by Lougheed's aggressive Conservatives and their demands for government action to create modern social services and a diversified economy. In 1971 the Socreds were swept out of power. The discrediting of Social Credit worked to the advantage of a Conservative dynasty. "You know, we could never have attacked that government as you did," Lougheed later told Reimer. "You really made it possible for us to get elected."

In May 1986, after 15 years of virtual one-party Conservative rule, Alberta New Democrats won 29 per cent of the popular vote and elected 16 members, 13 of them in Edmonton. Activists went wild celebrating their "sweet little sixteen," their coming of age as a major political force. Party leader Ray Martin paid tribute to those who had stuck it out in the dog days: to Grant Notley, his predecessor as leader, killed in a tragic air crash, and to Neil Reimer. It was the first public recognition of Reimer's leadership in almost 20 years.

There was something about the Socred years and the NDP's street-fighting style that people wanted to forget. "I think that people became frightened of you," Ted Chudyk told Reimer many years later. "You couldn't recover to become an Ed Schreyer again, the statesman. You were known too much as the two-fisted fighter. I guess maybe we got a little too deep in the mud." Then his mind flashed back. "Somebody had to do it. Everybody was waiting for somebody else to make them at least mortal for the first time, and to be subjected to public scrutiny like the rest of ordinary human beings. These guys never had been scrutinized before." They were a dynasty, he says, "well hidden behind the pulpit."

In a collection of essays honouring Grant Notley, political scientist Larry Pratt accused Reimer of losing sight of the new issues, the problems of white-collar workers, of mental health, education, and social services. "These were the issues of the day which the NDP neglected in its obsession with corruption in high places," Pratt writes in *Socialism and Democracy in Alberta*.

In fact, Reimer spoke to all those issues and featured them in all central campaign publications and statements. But he also understood the difference between what pollster Marc Zwelling calls the "contents" and the "container," the key to what's known as the "teflon" effect.

People may hate cutbacks, war-spending, and high taxes. They may love the environment, better hospitals, and better health and safety laws. But that never cost Ronald Reagan or Brian Mulroney an election and never won one for the NDP. People know the NDP stands for the average person; they just don't think the NDP can run a government. New Democrats don't like working on that "container" problem. They keep perfecting their content, watering it down or beefing it up, thinking that will make the difference.

Reimer suffered a "container" problem himself. No matter how reasonable his policies sounded, the street-fighter stuck in people's minds. He deliberately sacrificed respectability to go after the Socred "container," to use anger as a solvent against the major obstacle to political progress in Alberta. "If this party and its leadership cannot speak out against corruption, against the rape of this province, abuse of office," he told one rally, "then we will also be very unconvincing that we have the guts to establish the policies and programs that we want for Canada and Alberta."

He still thinks he was right. "We had to fight and there wasn't much to fight about in some respects economically, because the province was well off. We had to attack the integrity of the Social Credit. If we were going to allow this conservative element that draped itself in the Bible and couldn't play normal politics, we would never defeat them." Reimer understood Alberta Social Credit as only a Saskatchewan CCFer could. Both parties, both movements, started off as kissing cousins. Both were born in the Depression. Both hated greedy bankers and industrialists. Both hated lackey politicians in Ottawa. Both hated special privilege, insider deals, backroom boys. Both had deep fears about their society, their way of life, disintegrating. Both had inspired and religious leaders, visionaries and prophets. Both looked to major changes in the entire economic system.

These are the classic earmarks of populism. Both the CCF and Social Credit parties were populist, with one difference. The CCF was the real item. It was based on the grass roots. It was tied to independent organizations, co-ops and unions, centres of debate and action, kernels of a new order. Social Credit was always run from on high, directed from the centre, announced over the radio.

Reimer's swan song as leader was the sign-off of a socialist populist, a fighter. He was concerned with social breakdown, not just economic conflict. He was looking for ways to spread direct power, not just centralize government. He was not a statesman, not a social democrat -- he was more radical than that because at bottom he wanted people to free themselves in an open society.

Reimer borrowed his parting shot in his last speech from Kahil Gibran: "if it is a despot you would dethrone, see first that his throne erected within you is destroyed."

Garth Turcott, NDP-MLA, the lonely terrier among a pack of lapdogs.

NEW NDP LEADER NEIL REIMER
. . . receives congratulations from his wife

LEFT TO RIGHT: SHIRLEY CARR, LOUISE DION, JACQUES MALOFY, REG BASKEN, NEIL REIMER, ROBERT STEWART

A BICYCLE BUILT FOR 300,000

Fred and Louise Theobald got hooked on bicycles. In September 1981 they exchanged marriage vows after riding down the aisle on their two-wheelers. In 1984 they made another commitment. On May 20, they hopped on a bicycle built for two at the Crow's Nest Pass in Alberta and started an On to Ottawa trek. Their goal: to make it in time for the 1984 Energy and Chemical Workers' convention and along the way spread the word about the union's fundraising campaign for osteoporosis research.

As a marathon feat on behalf of a worthy cause, their jaunt doesn't match the long-distance runs of Terry Fox or Steve Fonyo. Unlike Rick Hansen, they skipped the gruelling Rockies and took a plane over northwest Ontario. As a political stunt to dramatize a new issue their jaunt lacked heroics. But the Theobalds didn't set out to rivet the country's imagination or set employers on their heels. They just wanted to do some legwork on an issue that posed the rethinking of age-old old-age stereotypes.

Theobald had been waiting for something like this since 1972, when he attended a seminar on how people look upon unions. He wanted to do something to correct that image, to show that unionism stood for more than organized greed, that it was a crusade for social betterment. The time, the commitment and the method of influence-pedalling were his.

The issue was also Reimer's. It came to him during a fit of rage after he saw two nasty letters to the editor of the *Edmonton Journal* in 1981. The first letter attacked unions for doing nothing on behalf of the elderly. A follow-up letter from a unionist agreed that unions got pensions for their members, then forgot about them. "I couldn't figure out why I was so mad," says Reimer. "Then I realized I was mad because it was true. Our view is: `now that we've looked after you economically, what more do you want?'"

Reimer contacted an old friend, Bill Cochrane, former deputy minister of health in Alberta, then chair of the Canadian Geriatric Association. Cochrane said that society was obsessed with youth, while medicine was obsessed with life-threatening diseases. Doctors were guilty of male-practice — focusing on injuries that affected men. Doctors were as out of touch with the needs of older women as they were with health and safety in the workplace, he said. They just doled out pills, instead of helping older women become self-reliant.

Reimer had some practical concerns. He wanted a project that could put the imprint of general social responsibility on the new all-Canadian Energy and Chemical Workers, while it was still young and impressionable. He wanted a project that a small union could identify with, manage, and measure progress on, not a billion-dollar sink-hole like cancer research. He wanted a project that had been overlooked by the big spenders, that could draw attention to political neglect. He wanted a symbol, but he wanted it to be tangible, to count for something, to let the union make a difference.

Cochrane told Reimer about osteoporosis, the "little old ladies' disease" that afflicts three hundred thousand women over age 45. Their bones become so brittle from loss of minerals that a sneeze can break them. Cochrane said that a donation of three-quarters of a million dollars could make a big dint in the work that had to be done.

Reimer intended to step down at the 1984 convention, and he liked the idea of osteoporosis research being his swan-song — a nice transition to retirement. The Theobalds volunteered to give the convention launch a running start with their bike tour. Their objectives were modest. They wanted a gimmick that would draw attention to osteoporosis as a preventable disease. They wanted to learn from senior citizens' efforts to stand tall, maintain their self-reliance and vitality. They wanted to stop over at companies represented by the Energy and Chemical Workers, to get the mainly male membership to take up a problem that affects so many Canadian women.

In 1984 these aims were a big deal. "Nobody could pronounce osteoporosis, let alone spell it," says Reg Basken. "Two years later, it

288

was a household word." More to the point, it became a household craze, milked for sales of another pill to pop. That's the way the cycle of issues works in North America -- the land, in H.L. Mencken's words, "of abounding quackeries."

By contrast, the Theobalds wanted to learn something that would stand the test of time about the social as well as physical side of osteoporosis. That's why they pushed their issue on a bicycle, not a bandwagon. Movements that endure build up speed slowly.

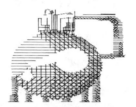

Prairie roads leave a lot to the imagination. Apart from thinking about his sore rear, Theobald thought about the vast and lonely distances they were travelling, got caught up in his own version of zen and the art of bicycle maintenance. In Ontario "The exterior attracts your attention, if only to fight the traffic. You've got to be on guard all the time, so you don't have time to internalize," he says. On the prairies, where bikers can use the extra lane built for tractors, there's just nature to worry about.

On the prairies "You have that feeling that you're always on the border of survival," Theobald says. "I think that's why wheat farmers plant green everywhere, to reassure themselves that there's some control, some environment they can create."

In the nursing homes they visited along the way, they found no space to create a personal environment. "I can see why old people stay old there," Theobald says. "It's a very constipated structure, very regimented. Like all institutions, they have a way of undoing themselves through the 'law of the worst.' To prevent the worst from happening, institutions develop rules that don't let people choose. So they end up segregating old people. It's the cheapest way to look after them."

Theobald had found his own reason for backing osteoporosis research: "To keep people independent — that's the hope to see in it."

After getting to Windsor they got to learn about life in the slow lane in Ontario and Quebec. As Theobald puts it: "Without a car, you're a nobody. No one has time for you, no one watches out for you, there's no place for you. Everything's organized on the assumption that everyone has a car. That excludes the old, the disabled and the poor. You can see the control that society has over people, without them knowing it."

By the time they got to the Ottawa convention the Theobalds had a bone to pick with the entire system of care for the aged, not just osteoporosis. "We were only the instrument of your understanding of the needs of older people," they told delegates. The first pledges were taken at the convention. The $750,000 goal was reached in 1989, partly financed by donations of two hours pay per union member. "It gave our people an opportunity to talk about the value of togetherness and commitment, something other than slot-machine unionism," says Reg Basken, who was elected director at the 1984 convention.

As the Theobalds were completing their road scholarship on aging, scientists with the U.S. National Institute of Health recommended calcium as the antidote to osteoporosis. Drug companies cashed in as North American sales of calcium pills soared from $18 million dollars in 1980 to $240 million in 1986. Most of the ads were pitched to seniors, who thought calcium was a cure.

In fact, calcium can only help prevent osteoporosis if it's taken by the young, whose bones are still growing. Even then, according to the May 1988 *Consumers' Report*, many of the calcium dosages are so poorly formulated that the body can't absorb them.

The calcium craze underlined the importance of non-profit, non-sensationalist, long-term research, such as that sponsored by the ECW. Four major sets of experiments were funded: studies in the bone differences of people with and without osteoporosis; studies on the beneficial impact of sodium fluoride treatment; studies into bone remodelling and mineralization; studies that showed the importance of diet and exercize in warding off the disease.

All of the studies were completed as a result of a commitment that financed a five-year, multi-disciplinary grant, a luxury most scientists don't enjoy, says Donna Metcalfe, executive director of the Canadian Geriatric Research Society. "That's the key to what the Energy and Chemical Workers have done. They allowed the researchers to get on with it."

GASOLINE ALLEY

Canada's oil and gas went down the pipe long before the free trade deal of 1990. Those resources have been mainlined to the United States since the 1950s, used to give U.S. industries a shot in the arm. That's when the possibility of using gas and oil as king-pins of a national industrial strategy was "tossed aside," writes Melissa Clark-Jones in *A Staple State: Canadian Industrial Resources in the Cold War*. Energy policy was "determined by American markets, by American government policies and above all by the internal interests of the major American oil companies."

The U.S. government looked to Alberta as its gasoline alley during the Korean War, when U.S. resources were stretched and the United States couldn't count on unstable Third World countries. When Alberta held up exports until its home market was served, U.S. diplomats enlisted C.D. Howe, federal minister of

industry, to lean on Nathan Tanner, Alberta's man in charge of oil and gas conservation. Tanner okayed pipelines to Washington.

Tanner turned up next as chief lobbyist for the Texans who won the contract for the Alberta-to-Ontario pipeline. A public line, or "anything that leads to statism," was intolerable to his clients, Tanner said in 1955. Then his clients demanded a low-interest government loan. To rush the loan through, Howe imposed closure on parliamentary debate on "Black Friday," 1956. Tagged by Tory John Diefenbaker as traitors to independence and democracy, agents of a "Colombo plan for Texas tycoons," the Liberals got the boot in 1957.

No oil company lost money underestimating Diefenbaker's nationalism. Henry Borden, Tory bagman, close friend of Howe's and chief of a major multinational, got a commission to develop a national oil policy. A royal commission on economic growth prepared by the outgoing Liberals' Walter Gordon said Alberta's oil was artificially "shut in" and had the capacity to serve the entire country. But Borden's report nixed a public and cross-Canada line, and greased the skids for more north-south trade. In return for a guaranteed market for its cheap Venezuelan crude in points east of Kingston, the United States agreed to take Alberta's extra gas and oil off its hands. Borden admitted his decision "involves considerations beyond the question of securing lower transportation costs."

Diefenbaker's government referred European orders for oil-based technology to the U.S. parent companies, Clark-Jones's research shows. In 1963, Canada and the U.S. signed a continental defence agreement to share "the continent's resources: oil,gas, electricity, strategic metals and minerals" and water.

POWER TO THE PEOPLE

Thinking isn't the only renewable resource in the energy and chemical industry, but it may be the rarest. Politicians jump at the prospect of megaprojects and tax grabs. Corporations leap at the chance of windfall profits to cover their front-end costs for research, exploration, and technology. Groups that lobby for environment or long-term jobs straggle along behind, cast as naysayers because they lack the hard edge that comes with insider information and connections.

But over the last 20 years, a vested interest has joined with public-interest groups. In 1969 the Oil, Chemical and Atomic Workers outlined a national energy policy. In 1984 the Energy and Chemical Workers laid out a plan for the chemical industry. They blew the whistle on their own employers. They put their jobs on the line to stand up for the environment and Native rights. They had the goods to call the government to account for forfeiting a national industrial strategy.

They weren't always so farsighted. Before 1960 the oil and chemical union was a

293

typical union, "mainly interested in getting a union organized," Neil Reimer says. Discretion was the better part of valour. "There's always a concern about attacking your own industry's social performance. Do you risk serious labour disputes, jobs, accusations of disloyalty to your own jurisdiction?" Upstart unions were supposed to know what side their bread was buttered on.

In the midst of the pipeline scandal of 1956, *Union News* carried a few articles echoing the CCF call for public ownership of a trans-Canada pipeline. But the union provided no expertise, no credibility with the media, and no rounded program to link pipelines with an overall energy or industrial strategy. While voters hailed Diefenbaker's bravado against U.S. ripoff artists, the April 1956 *Union News* ran a feature on "Prosperity in Canada." The article referred to the "mixed blessing and curse" of U.S. domination of the oil industry as Canada's "only handicap."

The union was too small, its resources too stretched, to participate in energy debates. It made no presentations to Walter Gordon's 1957 royal commission on economic prospects or to Henry Borden's 1958 royal commission on a national oil policy. Nor did it intervene in Nova Scotia, Ontario, or B.C government inquiries into shady practices at monopoly-run gas stations. This was the decade when foreign multinationals swallowed all the independent Canadian oil and gas firms. When Gulf bought out B.A., Reimer reacted like an organizer. He issued a cartoon to all Shell plants. No words, just a drawing of a big fish swallowing a school of little fish. The point was simple: The little fish had to get together.

Reimer didn't weigh in with a significant policy statement until 1960. Foreign corporations have taken over the industry, he told OCAW's Canadian Council. To reclaim Canadian control he called for royalty schemes to favour newcomers and co-ops and for export taxes to slow the drain of U.S. sales. To stake out public control, he called for government trusteeship of Alberta's tar sands, to be reserved for poor countries that had no oil.

He made no reference to the environment, or to the use of oil and gas as building blocs of a petrochemical industry.

"Hell, us stubble jumpers in Saskatchewan built an oil industry," Reimer says. "Why should we say we needed U.S. expertise? Seeing that need for an alternate expression in the industry stood me in good stead when we argued for Petro-Can."

The 1960 speech showed off the fancy footwork of a union bargainer fighting to keep a leg up on a powerful industry.

Reimer learned quickly that when he got fired up about oil and gas policy, "It was a good weapon in bargaining. The industry was defensive about its profile as a public utility, so I'd shoot in the direction that if they didn't behave themselves, we'd push harder on the other front to get rid of them."

When Reimer became leader of the Alberta NDP in 1962, he got tougher on the industry and himself. He needed details and full blown policies to stomp the government's record on the campaign trail. To subsidize this research, OCAW sponsored publication of James Freeman's *The Biggest Sellout in History*, a pioneer work of brass-knuckled investigative journalism. It predated the radical nationalist analysis of Jim Laxer, Mel Watkins, and Larry Pratt by a decade.

For five years Reimer matched wits with energy experts in public debate. He learned to think as the leader of an opposition party, as a strategic policy-maker responsible to the general public. He learned to speak to "the man on the street" about problems he knew about from the shop-floor or bargaining table. His work presents a case study of the NDP's role in raising the political and social level of union thinking. When Reimer resigned from the NDP in 1967 and returned as the Canadian director of OCAW, he quickly insisted on an energy policy that was "more clearly focused."

The 1970 meeting of OCAW's Canadian Council featured David Cass Beggs, former head of Sask Power, then in charge of Manitoba's power utility. Cass Beggs was open to a truly continental energy policy that included Latin America but warned that divvying up with the United States alone would be as equal as "the 50:50 pattern of the horsemeat in the chicken pies — one horse for one chicken." Canadian policy should stress conservation and long-term planning and should boost processing and energy-intensive industries, he said. The 1970 Council passed the union's first national energy policy. Public ownership of pipelines and reserves, upgraded domestic processing in Canada, ecology research, and northern control of northern development highlighted the 25-point program.

The union embraced Canadian nationalism, then considered a dirty word by corporations, Conservatives, Liberals, New Democrats, and unionists alike. "Canada is undergoing a struggle for independence," Reimer told the 1971 Council meeting. Continued ties to the United States will "consign our people to the permanent status of being an economic colony. We must regain the control of our economy, and in order to control the energy resources, Canada must use all the economic tools available to a nation."

In March 1973, half a year before the first oil shock of the decade, Reimer wrote Grant Notley, his successor as Alberta NDP leader, about the need to sharpen the NDP's tools. Instead of blanket nationalization, Reimer proposed a national energy corporation that could act as a watchdog on the industry and set the pace for environmentally-sound energy use. He put the same proposal to David Lewis and Tommy Douglas when they led a tour of the NDP caucus to

Alberta. Quite independently, the federal energy ministry broached the same idea in its July 1973 report, *An Energy Policy for Canada*.

Reimer says that Petro-Canada was, at first, "just a subsidy to help the industry." He says, "It was to lead in research and exploration. That was just a ruse, another way of giving a grant to cover the industry's up-front costs." In 1975 Petro-Canada was reborn with a $1.5 billion silver spoon in its mouth and a mandate to become a major player in the industry. It eventually bought out four foreign multinationals. Reimer remained unsatisfied. He wanted a buy-out of Imperial and the creation of a genuine energy corporation to oversee all facets of the industry.

In 1974 Reimer got the CLC to adopt OCAW's energy policy. He met with energy minister Marc Lalonde on behalf of the CLC.

Though it was a perfect time to jump on the energy crisis bandwagon and cash in on megaproject jobs, the union became the staunchest critic of crash programs. In 1973 Reimer gave Notley the inside dope on tar sands pollution — 130 tons of sulphur and 90 tons of fly ash per day — so he could pressure the government to slow down. In 1974 the OCAW's Canadian Council warned against the risks of environmentally dangerous projects.

In 1975 the Council invited Chief James Wah-Shee to deliver a keynote speech against the Mackenzie Valley pipeline and voted unanimously to support the Native land claim to block the project.

The Mackenzie Valley project was a $5 billion pipedream to carry Alaska gas to the United States. Trudeau appointed Judge Thomas Berger, formerly an OCAW lawyer and leader of the B.C. NDP, to review the social and environmental costs and economic feasibility of the project.

This isn't just about a pipeline, it's about the future of the north, Reimer told the Berger Commission in September 1976. Native culture is as delicate as its ecology, he said, and there would be only one chance to treat them right on the world's last frontier. Land claims and labour laws had to be in place before any development. Speaking on behalf of both the OCAW and CLC, he called for a halt on all megaprojects until a comprehensive energy and conservation policy was sorted out.

In 1980 the founding convention of the merged Energy and Chemical Workers adopted a 19-point energy program that stressed conservation, native rights, and the expansion of Petro-Canada. Canada is the worst energy waster in the world, Reimer said.

Though its members were in the business of producing energy for sale, the union took the view that limitation was the highest form of flattery. While every Canadian burns up 8.7 tons of oil a year, West Germans get by with 4.3 and the Japanese with 3, Reimer told the convention. Research on forms of renewable energy gets one-tenth the budget for research on oil and gas, and one-eightieth of that for nuclear, he said. Though the union represented atomic researchers in Ottawa, it called for shelving nuclear power until a public inquiry into health and safety, uranium waste disposal, and stepped-up conservation could be held.

Expansion of Petro-Canada was the union's top priority. "The stranglehold by large foreign-owned multinationals on our petroleum industry (and beyond, with their diversification into coal, uranium and other minerals) must be broken," the convention's resolution stated. Petro-Canada should be a full-fledged energy corporation that takes the lead in renewable energy.

The ECW backed most parts of Trudeau's 1980 National Energy Policy. The NEP was designed to stop the export of energy profits to the United States ($3.7 billion between 1975 and 1979), to strengthen Canadian firms, and to give Ottawa some dibs on oil royalties from the West. Petro-Canada was expanded. Reimer worked closely with Dr. Ed Clarke, the assistant deputy in charge, though they disagreed on two basic issues. The union opposed gas-price increases imposed by the NEP. Canada needs low energy costs to counter cold and distances, Reimer says. High prices just make pollution a class privilege, he says. He favoured strict standards on engine sizes and unleaded gas. The union also wanted the federal government to provide compensation to Western provinces, to make up for their sale of energy to the rest of the country at below-world prices. High natural-gas prices created a whiplash effect on the petrochemical industry, especially when coupled with Trudeau's 1981 deal with Alberta to levy taxes on gas used as feedstock for Canadian petrochemicals. In the aftermath, 27 plants worth $900 million were shut down. Sarnia alone lost two thousand jobs, and the industry lost 12 per cent of its capacity. The plants that survived were locked into huge debt loads left over from expansion in the 1970s — petrochemical investment went from $923 million in 1973 to $10.5 billion a decade later — and had to fight their way into a glutted world market by beating out competitors with lower feedstock costs. In 1983 the federal government appointed Reimer and Shell chief Len Bolger to head up a task force that would explore the industry's problems. Though a co-chair, Reimer was outnumbered thirteen to one by top industry people. In Sarnia the union set up a co-ordinating committee of all local executives to prepare a brief.

The petrochemical industry should be treated as "strategic," the union brief said. It accounts for a hundred thousand jobs in five provinces. It's on the leading edge of high technology. It pays good wages. It has strong world sales. Petrochemicals, not gas and oil, is where the jobs are. End-products multiply the value of raw materials 30 times, the union brief said. Yet only 5 per cent of oil and 8 per cent of natural gas gets processed in Canada. The rest goes up in exhaust, or goes out to the United States.

"The survival of the industry in Canada is really a function of government policy," the union brief said. Taxes accounted for half the cost of feedstocks. The brief called for lower taxes. It opposed government grants to save companies, except when the union was directly involved in decisions to improve productivity. The brief made no mention of environmental problems associated with petrochemical products. Nor did it raise concerns about the privatization of Polysar, via a sell-off of the crown corporation to the Canada Development Corporation investment pool.

The task force was "really taken aback" by the brief, Reimer says. "They'd never seen the union in this role, being that knowledgeable about the industry." When the task force reported in 1984, it supported the major proposals of the Sarnia unionists. That created a temporary high, Reimer says, and "members saw the union as very much with it, playing its role in their destiny, not only in terms of seniority and 15 cents an hour, but in the direction of their industry and community."

The task force recommendations were never acted on. The Conservatives, elected after the report was tabled, had no commitment to it. Although employers unanimously signed the report, Reimer says Imperial did so reluctantly, as a vertically integrated corporation that sells itself its own feedstocks. That gives Imperial the edge over its competitors who have to absorb the full price of feedstocks. Imperial, Sarnia union rep Stu Sullivan says, hopes to use its control of gas and oil "to control competitors like Polysar by the shorthairs." The free trade agreement forbids made-in-Canada prices that favour homegrown industries and makes impossible the union policy of using energy resources as the pivot for a national industrial strategy.

THE BIG CHILL

Until 1973 Canada had more oil and gas than it knew what to do with. After 1973 it had less gas and oil than it knew what to do with. The crisis turned energy politics outside in and upended relations between producers and consumers, west and east, Natives and whites, Liberals and corporations, energy savers and spendthrifts.

Until 1973, oil was an inflation fighter. The cost of living doubled from 1947 to 1971, figures from the Canadian Petroleum Institute show, but the price of crude oil stayed static. Canada had enough oil for 923 years and enough gas for 392 years, federal energy minister Joe Greene said in 1971. The June 1973 report of the ministry passed for a doomsday scenario. It predicted that energy costs might double or triple by 1990.

The ink was barely dry on that report when war broke out in the Middle East, and Arab countries scaled down their oil sales. By December 1973, oil prices overshot the predicted 1990 mark. By 1979 prices doubled again. By 1981, after two years of cutbacks from Iran, the price of crude was seven times the 1971 price.

Distress prices led to distress policy-making and financed a shopping spree for energy megaprojects. Canada will run out of conventional oil by 1984, the national energy board said in 1975. Sources had to be found in out-of-the-way places, rushed to market at costs that were out-of-bounds when gas and oil were dirt cheap. Whatever it took, oil could be squeezed from Alberta's tar sands and pumped from the ocean floor, and gas could be piped over the Mackenzie Valley tundra. Oil and gas reserves took priority over Native reservations.

The siege mentality ruptured federal-provincial relations. The new tilt in oil supplies meant Quebec and the Maritime provinces, cut off from Alberta pipelines, had to pay top dollar for world-priced imports. Ottawa had to subsidize the difference, or face the separatist music about Confederation being a shaft. If Ottawa had to pick up the tab for the East, it wanted its share of bonanza royalties from the West. The West, which had long paid top dollar for its imports, warmed to the idea of Eastern bastards freezing in the dark. It wanted the whole gate on royalties to put away for the rainy days when the oil boom ran out.

The oil shock brought relations between the federal government and the oil companies to a crunch. Lacking a majority, and with the NDP hard on his heels, Trudeau announced a new self-sufficient oil policy in December 1973. Oil prices were frozen. All Canada paid the same price. The pipeline was extended to Montreal. Sales to the U.S. were taxed. The green light was given to Northern pipelines and tar sands. And a national public petroleum corporation was formed. You've sold out to the NDP, Tories gasped. "We've gone beyond the Communist Manifesto," Trudeau shrugged. "Next question."

WHO HAS SEEN THE HOT AIR?

Neil Reimer was a male leader of a male union who dealt with male managers in a male industry. He's been married to Merry Reimer since 1945, counting on her to carry the daily burden of household chores while he clocked five million miles in air travel. He also counted on her to lead the union's and Alberta NDP's women's auxiliaries, and to leave the impression that he was his own man. But he created the least "macho" leadership style of any union in the country and relied heavily on her contribution.

Values, dialogue, and involvement stand out as a seamless web in Reimer's union style. They're in synch with what U.S. futurist John Naisbitt sees as a megatrend of the 1990s, as a rising generation of woman leaders brings focus to values rather than goals, dialogue rather than bravado.

Tommy Douglas, folk hero of the Saskatchewan CCF, became Reimer's role model. "I never saw a man so at home and so comfortable with his beliefs," Reimer says. "And I figured that's why he never felt compromised dealing with people from other walks of life, never tried to protect himself by staying in his own circle."

O.B. Males, Reimer's boss at the Co-op Refinery, taught him another lesson. According to Reimer, Males "had the discipline to divorce issues from personalities. It's key to what I call 'the art of disagreement.' That's more difficult and requires more skill than the art of just being an adversary. Anybody can go in and thump the table and shout."

Reimer had few opportunities to practise those skills in the 1950s and 1960s, when the union had to fight for survival and recognition. Those years explain his contempt for schemes to improve communication between union and corporate executives. Reimer says, "If it's only the union executive that has more rights to talk with management, it's counter-productive. People will only say the union's in bed with the company." The Oil, Chemical and Atomic Workers sat out the raging tripartism debate of the mid-1970s, when the CLC promoted three-way meetings among government, business, and union leaders. "You can't drive a car that doesn't have wheels," says Reimer. "They were talking about tripartism when we weren't even talking to one another."

Instead, he promoted dialogue between shop-floor workers and bosses around work organization. That freaked out union

UNIONISM WITH A HUMAN FACE

leaders who thought he had lost his class-struggle bearings, that the lines between workers and managers would be blurred. Reimer says, "I always figured that workers need a little less struggle, and a little more class. If a fellow digs a ditch all his life, his focus won't be blurred. But what future is there in that? "Society is more impersonal every day. If you're impersonal all day in your work, you're likely to be impersonal in your home and community. Human skills aren't increasing. Canadians don't talk to one another. Neighbours in the same apartment don't talk, let alone people from Alberta and Quebec. It is a great weakness in our culture. I thought we owed it to ourselves that we made damned sure we understood each other before we had a real confrontation."

Since retiring as director of the Energy and Chemical Workers in 1984, Neil Reimer has had a lot of time to think about the past and future of the labour movement.

He has kept busy as a delegate to the Edmonton Labour Council, vice-president of the Alberta Council on Aging, director of the Canadian Geriatric Research Society, a union nominee at labour arbitrations and the Unemployment Insurance Commission, a fundraiser for the inner-city Boyle-McCauley Health Clinic, and chair of his daughter's successful mayoralty campaign in Edmonton. He's an adjunct professor of workplace health and safety at the University of Alberta. He's a director of the leadership think-tank, the Niagara Institute, of the Edmonton Regional Airport Task Force, and of the Edmonton Community Foundation. He's a special advisor to the federal auditor-general.

He got to wear a tux for the first time in his life in 1989, when he was awarded the country's highest honour, the Order of Canada. He often asks why he's the only labour figure involved in most of these activities, how unions got to be on the margins of Canadian life.

Travelling with a well-heeled crowd also makes him realize that he can't kick the prairie dust of an immigrant sodbuster off his soles. On the day of the Order of Canada ceremony, he told the *Edmonton Journal*, "This is for my parents who kept the faith." As an individual, Reimer can't be understood except as a product of the plain folk who left the old world to settle on the Canadian prairies from 1900 to 1930. Being active in the community came with the territory.

As a unionist, Reimer can't be understood except as a product of the massive organizing drive of the Congress of Industrial Organization, the legendary CIO, which first brought unionism to heavy industry in the 1940s. In 1954 Reimer was the youngest of the CIO generation to lead a union. In 1984 he was the last of that warrior race to resign. He already felt out of place at CLC executive meetings. "As soon as the day's meetings were over, they'd rush off to their hotel rooms to make the phone calls they missed," he says. The camaraderie that went with struggle has tended to be replaced by the busy work that goes with bureaucracy, he says. Reimer received notice of his Order of Canada award, "for his work with founding the modern labour movement," while he was on a picket line outside the Alberta legislature protesting test-flights of U.S. cruise missiles over Canadian soil.

By definition, unions are about organization, not "the feeble strength of one." At their best, they are about "grass roots heroes." Only at their worst, or in newspaper headlines, are they about "labour bosses." Union members usually don't let their leaders get too far out of line with the kind of organization they want. But the collective nature of unions doesn't minimize the importance of individual leaders — it just explains the workplace and community culture that shaped those leaders.

Where leaders make a difference is in translating workplace problems and cultures into values, strategies, and issues. "Recognition of issues is the crucial leadership function in any union, whether it's an organizing drive, a bargaining campaign, or a political cause," Reimer says. Different leaders have different ways of defining issues, depending on whether they're stand-pat machine politicians or visionaries, whether they go with the crowd or strike out on their own.

Bob Stewart studied Reimer's style for many years, most of them as leader of a competing union. "Neil has been more than a builder of organizations," Stewart said at Reimer's retirement convention. "He has also provided much of the philosophy that governs the ways in which these organizations attempt to resolve the problems facing Canadians." Philosophy and Canadians are the operative words.

Old age may have given Reimer grace and knocked the kinks out of his hair but he admits that "friends like me now more than they used to. I was aggressive to the point of being a sonufabitch." Reimer says that a union leader can count himself lucky if he's respected, that popularity is not what leadership is about. But his sonufabitch years give us a close-up on the kinds of testy community and workplace traditions that mark Canada's unique labour history. They let us see some of Canada's problems from the unique perch of a scrappy labour leader.

"I guess there are very few jobs that allow you to talk to a janitor, a corporate executive, and cabinet minister on the same day," he says. That outlook has been put to good use not only in bargaining and in working in a number of social movements, but also as an entry into the broader problems of Canada's politics and identity.

Neil Reimer was born in Kitchkas, Ukraine in 1921, a year of famine and social breakdown after the revolution. His family, descendants of Dutch Mennonites who left their homeland in the sixteenth century to avoid the state religion, enjoyed high standing in the community. Reimer's father, Cornelius, commanded the local "safety army" which defended the town from roving bandits. Leon Trotsky, commander of the Red Army, was once a guest in their red brick house, attached, in Mennonite fashion, directly to the machine shed and barn.

But apart from perhaps being bounced on Trotsky's knees as a baby, Neil Reimer enjoyed no privileges. He was so undernourished that he was late walking and suffered permanent leg damage. Bandits made a point of terrorizing their home. Friends say that Reimer learned to talk himself out of tough spots by watching his mother convince robbers not to shoot him. He's never been rash, always looked for a fall-back, says his long-time colleague Reg Basken. "I think his family getting the shit kicked out of them in the Soviet Union made him think more than most of us," Basken says. In 1927 the family auctioned off their home and assorted goods and left for Canada. They rode a boxcar out of the Soviet Union and arrived in England, where they were stripped and deloused. Reimer's 14-year-old sister, Mary, had a puffy eye from a flying cinder and had to remain in England until it cleared up. "After all, we weren't looked upon very much more than cattle," she says. "And you don't import a sick cow."

The family travelled third-class steerage to Quebec City, then took a train to meet relatives in Milden, Saskatchewan. The four children and two parents lived in one room there until 1928 when they bought their own farm near Kindersley.

Only half of the land they bought had been farmed, or "broken," before they arrived. Their first winter was endured in a house without insulation. At school the kids were called "bohunks." At home they helped with farm chores and explored for Indian artifacts. In 1930, when the family was ready to celebrate its first full harvest, it hailed. "Everything was destroyed in a few minutes," says Reimer's sister, Mary. "It was kind of an empty feeling afterwards. We had an early start on the Depression with this hail."

For most of the next ten years, dust-storms were so thick, "You couldn't see the barn from the house," Reimer says. Grasshoppers and sawflies ate whatever grains survived after the topsoil blew away. Yet the price of beef fell so low that it didn't cover transport costs to the slaughterhouse. "This meant that Dad couldn't sell his calves, and he didn't have enough feed to raise them," Reimer says. "One day, with tears streaming down his cheeks, he went out to put an end to them." The family's workhorses died of starvation.

Depression stories from Saskatchewan sound like they come from the Book of Job in the Old Testament, or, worse still, from a broken record of parents' lectures to ungrateful children who don't know what it was like in the bad old days. Amazingly, there was no harvest of bitterness. Reimer says, "I often wondered whether our exchange had been a good one: from living beside the Dnieper River and the beautiful countryside, and moving to grasshoppers, cutworms, sawflies, dust gophers."

But on a summer Sunday in July 1929, Reimer's father called the family together, served up lemonade to everyone, and delivered a lecture. We are not going to complain, he said. This is our new home. Canada has given this land to us, and we all must give it something back, make a contribution. "It made an indelible impression on me," Reimer says.

Though Reimer was only six when he came to Canada, his immigrant background coloured his whole life. He had a weak sense of where he had come from, but a strong sense of where he had come to. He got used to asking about the "Canadian" meaning of the most humdrum aspects of life. "Being an immigrant meant I had to think everything through," he says.

The first person he saw on the train from Quebec was chewing gum. That was Canadian, he thought. He chewed over Canadian questions the rest of his life. "People want to know what kind of labour relations we should have," he says. "I say the first thing you should ask yourself is what kind of country do you want? Then you answer the question. If you just want labour relations for the purposes of making a profit, then the answer is simple. Get the buggers to work as cheap as possible and make sure they don't get a pension so they can starve to death afterwards. I think the question has to be asked all over again." Out of the mouths of immigrants.

Asking questions all over again was a central part of Reimer's Mennonite upbringing. Though Reimer's parents were "progressive" Mennonites, did not live within traditional communities or believe in the strictest doctrines of non-violence, they raised their family in core Mennonite beliefs. They weren't baptised at birth. That was something a grown person had to choose. Reimer says, "It was almost a religion in our family that individuals had to choose their own life, not parents, friends, or governments." He was raised to revere all life and even felt queasy about killing gophers. In Saskatchewan, gophers were so detested that the government put a bounty on them and paid schoolkids two cents a tail.

After high school, Reimer worked as a labourer for a year to save for a university education at Saskatoon's agricultural college. University president J.S. Thompson welcomed first-year students with a speech about university as a school for character, not jobs. Reimer still remembers one Thompson speech about education as a priceless endowment "to the extent that you have cultivated an honest mind, fearless in its love for freedom, inspired by a loyalty that is disciplined by fidelity to truth and the nobler ends of human existence." These were not easy words to live up to during World War II.

In 1941 Reimer was enrolled in compulsory officers' training. He refused orders to renounce his Mennonite heritage and the exemption from military duties granted to Mennonite settlers during the 1870s. He refused offers to accept special privileges as a Mennonite. He refused a physical exemption based on his flat feet and bad legs. He also refused to shoot at cardboard figures of enemy soldiers. He became Saskatchewan's leading refusnik.

"I just didn't think it had anything to do with university, and I didn't intend to shoot anybody," Reimer says. "They made it easy for me. They made it a human rights issue, which was supposed to be what the war was all about."

Forced to quit school, he packed his bags at the Hnatyshyn's boarding house (birthplace of the Canadian Governor General) and left for Regina.

Reimer started work at the Co-op Refinery in January 1942, just as a union organizing drive was getting underway. His workmates suspected that he was a management spy. Why else would a university-educated person work in a refinery? Reimer worked in the refinery because it was a co-op. His father was a founder of the fuel co-op in Kindersley. The year before Reimer went to university, he got most of his books from the co-op Wheat Pool library. He had come to see co-ops as the answer to Saskatchewan's problems.

"The capitalists weren't coming to Saskatchewan," he says. "They wanted our wheat and whatever natural resources we had so they could buy them cheap and ship them out. People saw first hand that the capitalist system did nothing for the people. They had to do it themselves."

The Co-op Refinery, which was established in 1935, was the world's first farmer-owned refinery. Plant workers organized a co-op cafeteria, bus service, and housing project for themselves. The Regina Labour Council sponsored a series of co-ops covering every human need. "I remember trying to sell a share in the labour council's co-op funeral home to one man, and he asked how he could collect his dividend," Reimer says. "These are difficult questions to answer."

The Co-op was a university in overalls. "It was like a school," Reimer says, because of all the spare time when things were going smoothly. "It was like a permanent union meeting, with every shade of opinion from Marxism to technocracy to social Catholicism."

Bill Ingram, one of Reimer's mentors, agrees. "We talked about everything, and of course the people on process — this is where we start to brag a little — were all pretty bright fellows. They all had at least grade 12 education," he says. "The socialism thing was under lots of discussion" and "we had every variety of opinion there was, even if there was only five of us on the shift," he says. "The nature of the thing was pure Saskatchewan."

Reimer quickly joined the union local, became assistant editor of the local's newsletter and then union president. The local's newsletter, the *ConCiLiator*, maintained a heavy emphasis on the human problems unearthed by industrial relations. One 1943 issue congratulated the co-op manager for "the good will he has shown in collaborating with the union in an endeavour to solve society's greatest problem of human relations." The Remembrance Day issue in 1943 warned, "If the terms of peace are signed on public documents, and not in the hearts of man, and in whose hearts hatred and war still reign, there will be no peace — not even for a day. We must abandon our unrestrained desire for pleasure, gain and to dominate others."

This concentration on human relations didn't reduce the differences between Co-op workers and the Co-op's farmer-owners. Nor did it eliminate conflict with the Co-op's management. Leaders of noble causes have as much difficulty recognizing the human needs of subordinates as fast-buck artists. Refinery manager O.B. Males was a towering Texan, gruff, intimidating, and with a crop of white hair to emphasize his senior years and wisdom. He wasn't politically correct, but the farmers who ran the Co-op hired him because he was a crack manager. He knew it, and he didn't like having his decisions questioned. "He would intimidate you, and you'd think he had apoplexy and was having a heart attack, " Reimer says.

When O.B. Males threatened to shut down the refinery rather than recognize the union, Reimer and his committee had their first union confrontation. "The good Lord came to my side and he put some words in my mouth," Reimer says. "Unless you recognize us in two hours, we're walking out and you can take the chance whether the men will be there with us," he stammered. "My knees were knocking and I didn't know whether they were going to come out with me or not," he says. Males gave in without a fight.

This was the direct-action style of the CIO, in the days before labour relations were enmeshed in regulations designed to keep outright confrontation to a minimum. The CIO is most easily remembered for its insistence that workers organize on an industrial, rather than craft, basis. But industrial organization was only one vehicle of the solidarity preached by CIO leaders. The CIO reached out to immigrant workers and minorities neglected by the craft unions of skilled workers. It endorsed the CCF as the political arm of labour, at a time when the CCF was running high at the polls, considered likely to form a post-war government. It practised militant shop-floor action — walkouts, slowdowns, sit-ins — to make its points with bosses.

Individual commitment, human brotherhood, and dignity on the job were the emotional buttons that CIO organizers pushed. Bargaining demands were simple, usually focused on union recognition and decent wages. Most strikes were fought for wage levels that barely brought workers above the poverty line of the day.

The Canadian Congress of Labour that chartered the Co-op local in 1942 was CIO through and through. So was the Oil Workers International Union that Co-op workers joined in 1948. In the days before unions collected dues impersonally through an automatic dues check-off, joining the union was a measure of deep personal commitment and belonging. The Oil Workers had a secret ritual and formal oath for new members, along the lines of traditional fraternal societies. Adopted in 1951, the oath made each member swear never to discriminate on the basis of colour or nationality, to defend freedom of thought, and to educate people about how capital and profit were taken from labour.

Unions had a "mission," the pledge said, "to fully secure the toilers' disenthrallment from every species of injustice" and to "subordinate every selfish impulse to the task of elevating the material, intellectual and moral condition of the entire labouring class."

Unlike craft unions in the American Federation of Labor, the CIO declared itself a movement, and the movement was bigger than its collective bargaining parts. Many of its goals couldn't be bargained. "We weren't thinking in terms of two cars and a boat, but respect, dignity, and the basics of food and shelter," Reimer says. "I told our members many times that if all they want from me is help in collective bargaining, I'm not interested. That's removing the movement aspect of labour unions. A movement of what? A movement of getting more money?"

In 1961 Reimer warned the businesslike unions in the Alberta Federation of Labour, "A labour union devoted to nickels and dimes is a union that will die." The challenge for unions is to preserve individuality in an age when automation threatens to turn people into slaves of machines, he said.

Though the CIO was founded in the United States, it died there too. CIO unions weren't as big as the unions in the AFL. When the two organizations merged into the AFL-CIO, the AFL came out on top. AFL leader George Meany, who took pride in the fact he never led a strike, led the AFL-CIO from its formation in 1956 until his death in 1980. The CIO didn't merge with the AFL, it was absorbed.

"We adopted the business union philosophy as a mentality, and it suffocated our spirit," says Tony Mazzochi, secretary-treasurer of the Oil, Chemical and Atomic Workers and a long-time dissident in U.S. union circles. "No one was willing to rock the boat, from foreign policy to politics. So it was just a whole move to respectability and conformity, to play by the rules that were established by a conservative AFL leader," he says.

"The opposite happened in Canada, and obviously they're still a viable movement up there as a result," Mazzochi says. In Canada, the merged labour body was called the Canadian Labour Congress to banish AFL-CIO differences to the past. The merger convention called for independent labour political action. Although the CLC's first leader, Claude Jodoin, came from a conservative craft union, all other CLC leaders came from industrial or public-sector unions. Most of the differences between Canadian and U.S. unions to this day can be traced to the fate of CIO ideals in both countries.

Neil Reimer remained the leader of one union for so long because he lost in his run for the CLC secretary-treasurer post in 1974. Because of this defeat Reimer remained a hands-on union leader past his years of frantic energy and mid-age crisis, into his years of calm, confidence, and wisdom. Because he was kept out of the CLC club, "The only route left to him has been to develop his union in the best way he knows," James Bagnall wrote in the *Financial Post* in August 1980. "For energy and chemical workers, that's probably a major gain."

His last decade was probably the most fruitful of his career. That's when he led the drive for a merger of energy and chemical workers in a Canadian union, for a Canadian energy and petrochemical policy, for a reassessment of Quebec's role in a national union, and for industrial democracy (or Quality of Working Life).

He retired in 1984, before reaching the age of compulsory retirement. His resignation letter to the national executive board in May 1984 gave three reasons for the decision. The union needed a new leader to keep a step ahead of the times. The union was in good shape and could run on its own momentum, giving a new leader time to grab hold. And he didn't want to squander the good will towards him by hanging around another term.

In his retirement speech at the 1984 convention Reimer talked about how the information revolution had increased the amount of information by 1600 per cent since he started with the union 40 years earlier. At that rate, he told delegates, "You will have to go through the changes in the next ten years that I went though in my lifetime." To commemorate his retirement, ECW staff established the Neil Reimer Leadership Education and Development Fund to prepare rising leaders for community and workplace leadership. "If any particular theme threaded itself throughout my career, it was worker involvement in our society," Reimer said in his farewell thank-you address.

THE REST IS HISTORY

AN AFTERWORD BY THE UNION EXECUTIVE

It's hard to write a union history with a happy ending. It's good to write a history that reminds us of where we come from, reminds us that what we sometimes take as given was given to us by old-timers who slugged it out in tough strikes and long campaigns. But it would be wrong to write a feel-good history that implies we made no mistakes, that we can rest on our laurels, that we're chugging along smoothly.

In the labour movement we can't just say about history that "what's done is done." We have to keep on doing it all over again. Despite all the progress we've made since the Depression of the 1930s, many of the problems that unions tried to deal with in the bad old days are still with us. There's no drop-off in the numbers dying at work, for instance. Some problems — the environment is only one example — have even got worse.

In 1984, when Neil Reimer retired after 30 years of leading the union, we wanted to take advantage of his long record to review the lessons of our history. We decided to sponsor research into our history and to give free rein to an outsider to analyse it because we wanted something to guide us in the future.

The book shows that the identification of new issues is both the hallmark of the union and the key to its greatest successes. The union wasn't just defensive, didn't just react, didn't just count on knee-jerk responses. It was open to taking risks, to asking the members for direction, to appealing to the community for help. It was prepared to "go for broke" to establish breakthroughs.

It's too easy to forget this formula in the course of the daily grind of union work. But it would be very dangerous to go into the 1990s without that kind of approach.

The members of our union are at the centre of most of the major changes sweeping the world. Many of us work for energy and chemical companies that pollute and destroy the environment. We work in industries that are movers and shakers in the new stateless economy that lets big business prowl the world in search of a better deal, with no regard for communities or national boundaries.

We work for companies that are highly automated, that tend to treat individuals as cogs in a machine. We're one of the few private-sector unions with members spread across the entire country, at a time when the country seems to be splitting apart, without much sense of common ground or even dialogue.

Those are the big issues we have to tackle. Otherwise, we'll be on the back end of change. We can learn a lot from this history about how to put ourselves at the front end.

We can learn to start from basics, to ask the big questions about democracy, involvement, and meaning on the shop and office floor and in the community. We can learn the importance of dialogue and the need for commitment to open discussion and to work difficulties through. We can learn the importance of throwing down the gauntlet when that's necessary to make headway. We can learn that we make the greatest progress when we work with others on broad issues of community importance.

The rest is history. It's good to come from a union that leaves such strong foundations for the future.

R.C. (Reg) Basken,
President

F.N. Gagné,
Secrétaire Trésorier

R. Boucher,
Québec

J. Haley,
Member at Large

J.R. Healey,
Ontario

W. Kolba,
Alberta and B.C.

B. Risling,
Member at Large

G. Steininger,
Manitoba and Saskatchewan

R.A. White,
Member at Large

R.P. White,
Atlantic